W9-BER-358

Painting Fabulous Floorcloths

Painting
Fabulous
Floorcloths

Sterling Publishing Co., Inc. New York
A Sterling / Chapelle Book

Chapelle, Ltd.:
- Owner: Jo Packham
- Editor: Laura Best
- Staff: Marie Barber, Ann Bear, Areta Bingham, Kass Burchett, Rebecca Christensen, Dana Durney, Holly Fuller, Marilyn Goff, Holly Hollingsworth, Susan Jorgensen, Barbara Milburn, Linda Orton, Karmen Quinney, Leslie Ridenour, Cindy Stoeckl

Plaid Enterprises:
- Editor: Mickey Baskett
- Staff: Jeff Herr, Laney McClure, Susan Mickey, Dianne Miller, Jerry Mucklow, Phyllis Mueller

If you have any questions or comments, please contact:
Chapelle, Ltd., Inc., P.O. Box 9252, Ogden, UT 84409
(801) 621-2777 • (801) 621-2788 Fax • chapelle1@aol.com

Library of Congress Cataloging-in-Publication Data

Painting floorcloths / Plaid.
    p.   cm.
    "A Sterling/Chapelle book."
    Includes index.
    ISBN 0-8069-6521-5
    1. Painting. 2. Floor coverings. 3. Stencil work. 4. Rubber stamp printing.   I. Plaid Enterprises.
TT386.P38   1999
746.6–dc21   99-21383
                                  CIP

10 9 8 7 6 5 4 3 2

A Sterling/Chapelle Book

First paperback edition published in 2000 by
Sterling Publishing Company, Inc.
387 Park Avenue South, New York, N.Y. 10016
Produced by Chapelle Ltd.
P.O. Box 9252, Newgate Station, Ogden, Utah 84409
Originally published in hard cover as Painting Floorcloths
© 1999 by Chapelle Limited
Distributed in Canada by Sterling Publishing
% Canadian Manda Group, One Atlantic Avenue, Suite 105
Toronto, Ontario, Canada M6K 3E7
Distributed in Great Britain and Europe by Cassell PLC
Wellington House, 125 Strand, London WC2R 0BB, England
Distributed in Australia by Capricorn Link (Australia) Pty Ltd.
P.O. Box 6651, Baulkham Hills, Business Centre, NSW 2153, Australia
Printed in China
All rights reserved

Sterling ISBN 0-8069-6521-5 Trade
           0-8069-6607-6 Paper

The written instructions, photographs, designs, diagrams, and projects in this volume are intended for the personal use of the reader and may be reproduced for that purpose only. Any other use, especially commercial use, is forbidden under law without the written permission of the copyright holder.

Every effort has been made to ensure that all information in this book is accurate. However, due to differing conditions, tools, and individual skills, the publisher cannot be responsible for any injuries, losses, and/or other damages which may result from the use of the information in this book.

Due to the limited amount of space available, we must print our patterns at a reduced size in order to give our patrons the maximum number of patterns possible in our publications. We believe the quality and quantity of our patterns will compensate for any inconvenience this may cause.

Introduction

WHAT IS A FLOORCLOTH?

A floorcloth is a decorative floor covering made of heavy canvas cloth or vinyl. Floorcloths are not new, but rather they are a revival of trends and ideas from the past, using different color combinations and techniques to go with today's decorating styles.

Floorcloths reached their initial peak of popularity in Europe in the mid-eighteenth century. They were considered an elegant decorating accent, while offering a functional floor covering.

In America, floorcloths were the forerunner of linoleum and factory-made floor coverings. Coming from the elaborate homes in Europe, the early settlers in North America, as they began to move west, were struggling to add a touch of refinement to their rough-hewn houses with dirt floors. Pattern and design were important. Initially, it was achieved by sweeping geometric patterns into the hard, clay dirt. Next came simple wooden floors that were generally spattered with paint. The spatter painting gave way to large geometric designs, generally black and white.

Enter the era of the imaginative housewife. As an economic measure, these domestic engineers opted to paint repeated patterns, via stencils, on sailcloth. The early floorcloths were base-coated with milk paints, sewn together, then stretched taut. They helped to block out cold drafts and could be moved from room to room. In 1796, George Washington listed a floorcloth for $14.82 in his financial disclosure report, and listed it as a possession he wanted to move with him. This was indeed a valuable possession in 1796.

Floorcloths in the twentieth century are again making a strong decorating statement. They are no longer an alternative to another type of flooring, but rather are viewed as an elegant decorating choice. Floorcloths are now offered by interior design studios as an ultimate finishing touch.

Besides their elegance, floorcloths still remain very functional. They are easy to maintain, very practical to use in high-traffic areas, and most importantly, are an affordable means of visually tying a room together. A definite plus for the do-it-yourself decorator is the availability of quality, preprimed canvas and vinyl remnants, stencils, stamps, paints, and nonyellowing varnishes. With the instructions given in this book, plus the available supplies, you will be able to create intriguing, decorative, and functional floorcloths to enhance your decor.

WHERE NOT TO USE A FLOORCLOTH

Floorcloths may not be the perfect floor covering for all areas. Consider these factors:

• A floorcloth can be slippery if placed on a highly polished or glazed tile floor. They can also be dangerous if placed on waxed, wood, or stone floors at the base of stairways.

• If a floorcloth is placed on a tiled floor with distinct grout lines, it will droop into grout lines and the finish could crack.

• A floorcloth should not be placed on carpeting with a thick rubber padding. Thin heels may break through floorcloth and cause irreparable damage.

Contents

Contents

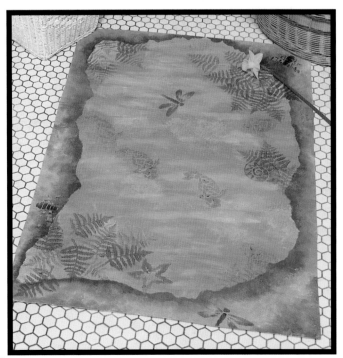

General Instructions

FLOORCLOTH DECORATING TECHNIQUES

There are many wonderful yet easy techniques to decorate floorcloths. Here is an overview of techniques covered in chapters to follow:

Blocking:

Blocking is an age-old craft that has found new life in home decorating. Traditionally, blocks were cut from wood, then later from linoleum. The cut block design is coated with paint or ink, then "printed" or pressed onto a surface. Blocking typically gives a solid, filled-in shape. Today, this process is made easy for the do-it-yourself home decorator with many innovative, easy-to-use products.

Decorative Painting:

Decorative painting combines the use of patterns with the painter's own freehand abilities. Generally, there are no special tools needed, other than paints and brushes. Highlighting and shading set decorative painting apart from other forms of painting, resulting in a more detailed and dimensional look.

Découpaging:

Découpaging is the art of decorating a surface with cutouts. Using a découpage finish, which is a water-based glue and sealer in one, fabrics or paper can be adhered onto a floorcloth to create beautiful designs and patterns.

8

Faux-Finishing:

Faux-finishing encompasses a wide variety of techniques, all of which have the goal of re-creating with paint, the surfaces found in nature. Some of these finishes include marble, stone, wood, leather, crackling, and tortoiseshell. Not only is faux finishing a less-expensive option to purchasing the real thing, it gives the ability to impart an age-old look to a surface, which normally would take decades to achieve naturally.

Stamping:

Stamping has come a long way since the days when carved wood and vegetables were used to create patterns. Stamps are now made of a dense, pliable material, which makes it easier to apply to a variety of surfaces. Stamping creates a shape, which includes some interior detail, such as the veins of a leaf. Today, stamping is easy to do for the do-it-yourself home decorator. Innovative products are available, creating a crisp design and making cleanup easy.

Stenciling:

Stenciling is an age-old, decorative technique of applying paint to a surface through the cut-out areas of a stiff, paint-resistant material. In today's market, stencils in every pattern imaginable are available, from the simplest piece of fruit to the most elegant Renaissance scrollwork. Not only are stencils easy to use, they do not require a lot of additional supplies—a few brushes and paints. The result of this easy-to-repeat method is designs with crisp, clean edges.

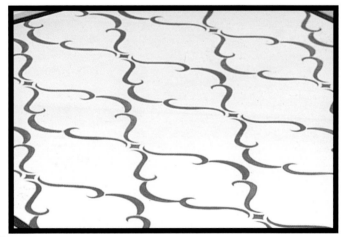

USING VINYL REMNANTS

A surface for floorcloths that is very affordable and requires no hemming is the backside of a vinyl flooring remnant. These remnants can be cut to any size and offer a nonskid, protective surface for the flooring underneath. Vinyl remnants are available at local home design or floorcovering stores. Because painting is done on the back side of the remnant, the front design does not matter. However, make certain to select a piece of vinyl remnant with an adhesive or paper backing, or a smooth, shiny, vinyl backing.

USING CANVAS FOR A FLOORCLOTH

There are two types of canvas used for making floorcloths:

• Preprimed canvas has been coated with protective layers of gesso and has had its selvages removed to greatly reduce the chances of shrinking and rippling. Some arts and crafts or home design stores sell preprimed canvas in precut, standard rug sizes or in rolls. This type of canvas is recommended for the beginning floorcloth maker.

• Unprimed canvas, or raw canvas, is available from awning companies or companies producing boat covers, as well as art supply stores. It is available in a variety of sizes and weights. The weight of the canvas is graded by number; the higher the number, the lighter the canvas. For example, a #5 weight canvas is much heavier than a #8 weight canvas. For priming and painting an unprimed piece of canvas, stretch and tack the canvas tightly to a stretcher or on a flat surface, using a staple gun.

To make large, room-sized floorcloths, canvas pieces can be sewn together, using a heavy-duty sewing machine and waxed thread or sailmaker's twine. Open out the seam and glue it down on each side with a heavy-duty adhesive.

BUILDING & USING A STRETCHER

A stretcher can be constructed for unprimed canvas floorcloths, using firring strips:

1. Purchase firring strips (rough or second-grade lumber is fine) at least the finished size of the cloth.

2. For corners, use metal L-clamps, which can be attached with wood screws, eliminating the need to drill and glue edges. The frame is easy to disassemble and readjust for another size of cloth by simply removing the L-clamps.

3. Remove wrinkles from canvas with a dry iron and spray starch. Some wrinkles will be removed by stretching, but sharp fold lines need to be removed before stretching and painting.

4. Before attaching canvas to frame, unroll on a large floor area. Measure and mark finished size of cloth, then add sufficient amount to turn under for hem, if hem is desired. Add another 1" to 2" for attaching to frame.

5. With finished size marked on canvas, it is easier to attach to the frame by pulling and stapling canvas from center to corner, then pulling opposite corners until cloth is taut. Continue to add staples around perimeter every 6" to 8", stretching and stapling in each section.

6. For larger cloths, attach support pieces horizontally on underside of frame every 2' to 3' to prevent canvas from drooping into frame.

7. Attach unprimed canvas to frame, working from center to opposite corners, stretching tightly. Do not worry if canvas is not perfectly square on frame. Finished size and hem markings will help when pulling from center to edge as evenly as possible. After painting, floorcloth will be squared and trimmed before hemming.

General Supplies

FLOORCLOTH PAINTS

Any of the following paints can be used for base-coating floorcloth as well as painting designs.

Laytex Wall Paints:

Latex wall paints are available in a multitude of colors and wear well over time. They are an excellent primer or base coat for canvas or vinyl. However, they are not recommended for design painting because the color is not concentrated enough and consistency is too thin for successful brush strokes.

Acrylic Craft Paints:

Acrylic craft paints are premixed and quick drying, and available in an extensive variety of colors. Acrylic craft paints are excellent for decorative painting as well as stenciling. They are also easily blended to create custom colors to coordinate with home decor. Cleanup is easy with soap and water.

Indoor/Outdoor Acrylic Gloss Enamels:

Indoor/outdoor acrylic gloss enamels are versatile and durable high-gloss enamels which wear wonderfully inside, but are weather resistant enough to go outdoors as well. They come in a variety of colors, adhere well to slick surfaces, and can be used for decorative painting as well as stenciling. These paints are available in most art and craft stores, and home decorating centers.

Indoor/Outdoor Satin Acrylics:

Indoor/outdoor satin acrylics are available in a wide variety of beautiful colors with a satin finish. These paints are not only durable on surfaces which receive a lot of wear, such as floors, but they stand up well to the weather. They are preferred for projects that will be used outdoors. No sealing is needed for these sturdy, water-based paints. They make cleanup easy with soap and water.

Stencil Paints:

Stenciling paints include a wide assortment of paints that are suitable for stenciling floorcloths.

Faux-finishing Glazes:

Faux-finishing glazes are rich, thick mediums with a subtle transparency that are perfect for stamping, block printing, and faux finishing. They are available in a wide variety of decorator colors, water-based, and nontoxic. Cleanup is simple with soap and water. Neutral glaze can be mixed with glaze colors to create an assortment of tints for textured effects.

Other Supplies:

There are some basic tools which, while not absolutely necessary, will make creating floorcloths much easier. You will probably find additional tools to add to this list. Remember, it is not necessary to have all of these items in order to begin:

• Carpenter's square—18" to 24" size, for use as a straightedge or to measure a 90° angle. *Note: If any measurement tools are metal, they need to have a backing so as not mark painted surfaces. Attach pieces of Velcro® (use the softer half). It comes with peel-off, self-adhesive backing so it is easy to attach several pieces on the edges of the ruler. The slight elevation to the straightedge allows for striping with a permanent marking pen with no underbleed on the surface.*

• Chalk line—for squaring up and marking grids on vinyl or canvas for a larger floorcloth.

• Craft knife (pencil-style)—for mitering corners and trimming edges of fabric appliques.

• Disposable plates and mixing sticks—for palettes when using and blending acrylic paints and mediums.

• Extrafine-grit wet/dry sandpaper—for sanding between base coats and finish coats. An alternative is a crumpled brown paper bag.

• Heavy-duty kitchen scissors—for trimming canvas and vinyl.

• Light blue or gray chalk pencil—for marking on canvas and vinyl. These marks are easy to remove with a white artist's eraser.

• Masking tape—for striping and edges. Use four widths: ¼", ½", ¾" or 1".

• Paint rollers—for base-coating and adding finish coats. A 2" disposable trim roller and tray can be washed and used several times. Use one for base coats and another for finish coats. Rinse roller tray and wrap roller in plastic wrap between coats. Replacement roller covers are also available.

• Painter's tape—for striping on prepainted surfaces. Painter's tape has one "sticky" edge and it is recommended in areas of high humidity or where base paints may be lifted with masking tape.

• Permanent marking pens—for adding shading to striping. Use a marker with a wide tip. A soft gray tone is best.

• Plastic wrap—for use in faux-finish process for "ragging off" background glaze.

• Quilter's ruler—for measuring grid increments and inside striping. Use a see-through ruler that is 2" wide and marked with a grid.

• Rolling pin—for "hemming" floorcloth. Cover rolling pin with aluminum foil, which protects surface from glue and makes cleanup easier.

• Sea sponge—for sponging on faux-finish backgrounds.

• Sponge brushes—for painting on striping (1"), and for base-coating and finish-coating (2").

• Tack cloth—for removing residue from floorcloth after sanding and before and between finish coats.

• Tape measure or yardstick—for measuring exact center or diameter.

• White artist's eraser—for removing chalk pencil markings or to remove a paint smudge when stenciling.

VARNISHES & SEALERS

Acrylic Sealers:

Acrylic sealers protect floorcloths indoors or outdoors and come in either a matte or gloss finish. They brush on and dry with no yellowing.

Polyurethane Outdoor Varnish:

Polyurethane outdoor varnish protects floorcloths with a durable finish that will stand up to the weather of outdoors. It brushes on and dries clear with no yellowing. It is available in a matte, gloss, or satin finish.

Spray Acrylic Sealers:

Spray acrylic sealers protect floorcloths for use indoors. They spray on with no yellowing and are available in a lacquer-like high-gloss finish, a subtle glossy sheen, or a soft matte finish.

14

Preparation

PREPARING A FLOORCLOTH

Priming Raw Canvas:

The importance of primer and base coat cannot be maximized. The primer is needed to level the weave of the canvas and to provide an even painting surface. It protects canvas from wear, yet remains flexible to prevent cracking. Gesso is by far the best option because the formulation and consistency are correct for sealing raw canvas and keeping it flexible. Gesso is available in a selection of colors, meaning fewer base coats are required.

If you are working on unprimed canvas and do not provide an even, smooth surface in this initial step, your error will not be realized until the final finish stage when the varnish soaks and discolors the floorcloth. At this point, there is nothing that can be done to recover the many hours of time and the material expense. The only remedy is to start over with another piece of canvas.

If gesso is not available, exterior latex paints can be used. Exterior paints remain more flexible than interior paints. It is not necessary to purchase the most expensive grade. Latex paints are recommended over oil paints because of the ease in clean up and to maintain compatibility of base coats to finish coats. For stenciling paints, there is not a problem with compatibility as with finish coats. Either paint crayons or acrylic-based paints can be used.

1. Apply two coats of gesso with a low-nap roller. Allow adequate drying time between coats. After second coat, lightly sand canvas with extrafine-grit sandpaper or a crumpled brown paper bag. Wipe surface with a tack cloth to remove any residue.

Base-coating Canvas:

1. Apply two thin coats of acrylic or latex paint in the appropriate color with a low-nap roller to cover floorcloth. Allow adequate drying time between coats.

2. After second coat, lightly sand with a crumpled brown paper bag. If using interior latex paint, thin paint with some water (seven parts paint to one part water).

For smaller floorcloths, use a trim roller (available in packages which include a disposable roller tray). If using a sponge brush, use a brush at least 2" wide and dip brush in water first, then blot on a towel before painting. This allows for a more even paint flow.

Do not rush drying time. A good way to test dryness is to take a "temperature" reading of the canvas' surface. If it feels cool to the touch, then it is not completely dry.

If painting in extremely humid conditions, more rippling of canvas will occur. A fan or dehumidifier will be helpful to improve drying conditions.

Priming Vinyl:

1. Cut vinyl to size. Prime back side of vinyl with a bonder/primer or stain-blocking primer.

2. Paint primed surface with latex wall paint in desired color.

3. Create background textures (sponging, ragging, etc.) and/or borders with paint or faux-finishing glazes, according to project instructions.

4. Create paper proofs to lay out design and plan corners and spacing.

5. Apply surface designs, according to project instructions. Allow designs to cure 24 to 36 hours.

6. Apply three to five coats of a water-based topcoat sealer. Roll or brush on and allow adequate drying time between coats. Rub surface with crumpled brown paper bag after third coat. Wipe surface with a tack cloth, then apply another coat.

CLEANING & STORING

To store or move, wipe clean with damp cloth, then roll floorcloth with decorated side on outside. (This helps floorcloth lay flatter.) If floor-cloth begins to show wear, clean surface and apply additional coats of same water-based topcoat.

MEASURING & CUTTING

When canvas or vinyl is unrolled and is exactly the size you want, measurement is easy; however, that is generally not the case. Mark surface with a light-colored chalk pencil. On a very large piece, a plumb line, or chalk line (with light-colored chalk), is the best measuring tool. When using a metal straight-edge, such as a carpenter's square or ruler, be certain to attach Velcro® or cork pieces to the back so it does not mark surface. Cut canvas or vinyl with utility knife or carpet knife.

Measuring & Marking a Rectangular or Square Floorcloth:

A floorcloth must be "squared off." To accomplish this, all measurements must start from the center point.

1. Refer to Figure 1. Using a tape measure or yardstick, measure each side of floorcloth and divide each side in half; mark halfway point on each side.

2. With a light, broken line, connect sides.

3. Refer to Figure 2 on page 17. Use a right angle or carpenter's square to be certain center lines are at right angles to each other.

4. Connect diagonally from corner to corner. The point where the diagonal lines intersect is exact center of floorcloth.

5. If necessary, trim outside edges and check outside corners with a carpenter's square.

Measuring & Marking Circular Floorcloth:
1. Refer to Measuring & Marking a Rectangular or Square Floorcloth. Measure and mark exact center of floorcloth.

2. Determine diameter of floorcloth. If diameter (distance from one side of circle to opposite side) is 60", cut a piece of canvas or vinyl that measures 60" x 60".

3. Refer to Figure 3 on page 17. Tie string to a pushpin and secure pin on center mark of floorcloth.

4. With pushpin in place, measure string to half diameter of circle (this is the radius).

5. Refer to Figure 4 on page 17. Tie pencil to string so pencil point is the full radius from pushpin.

6. Bring pencil point to end of string and carefully draw pencil around, keeping string stretched taut and pencil vertical.

7. Cut circle with sharp utility scissors.

Figure 1

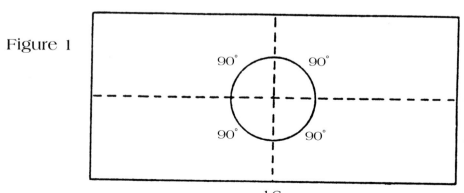

Figure 2

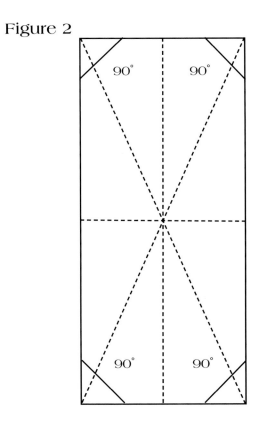

Figure 3

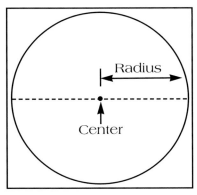

Figure 4

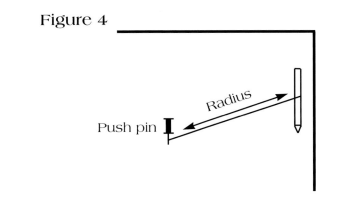

CAUSING A FLOORCLOTH TO LIE FLAT

For aesthetic purposes as well as those of safety, the edges of the finished floorcloth should lie flat. Heavier canvas pieces or canvas that has been primed on the front and back will lie flatter. If edges curl up, hemming is required to keep it flat as well as making it safe to have on the floor. If, after hemming, the floorcloth still does not lie flat, there are more options:

• Place floorcloth on a smooth, flat surface (in an area as humidity-free as possible). A concrete or carpeted basement floor holds more moisture and is not recommended. If no other surface is available, run a dehumidifier for several days in the area where the floorcloth is placed.

• Weight edges with books, covered bricks, or other heavy articles to stretch and hold down edges.

• Purchase firring strips and attach floorcloth to frame, using very fine wire nails.

• Place floorcloth on thin jute pad, cut pad slightly smaller than finished floorcloth.

Decorative Elements

CREATING LAYOUT

How do you know where to begin on a floorcloth, and how do you know when you have too much or too little for a balanced floorcloth design? Planning an overall floorcloth is similar in many aspects to planning for overall wall stenciling. If this is your first floorcloth, it would be worthwhile to plan the design first on practice paper. It is not necessary to mark off the finished size on practice paper. If the finished size is to be 4' x 6', cut a piece of paper that is ¼ the size. For example, for a 4' x 6' floorcloth, cut a piece measuring 2' x 3', then cut this piece into a triangle. This will let you plan a corner, a border, and a portion of the center pattern.

Before placing any design elements, double-check measurements. Be certain corners are square and floorcloth is marked for a finished edge. If a hem will be added, make allowances.

RANDOM DESIGN

Random design is simply the absence of repeated motifs. Because of its free-form nature, there are no set rules; the only require-ment is that the design be aesthetically pleasing to you. A great example of possibilities of random design are seen in the project Garden Reflecting Pond on page 116.

OVERALL DESIGN

You may want to place particular design motifs in a symmetrical pattern that covers the entire floorcloth. To plan placement of these motifs, first stencil several design motif elements on practice paper, cut them out, then position them to determine spacing in the area available. Once the number of motifs that will "work" in each row is determined, follow these steps for measurement:

If All Motifs Are the Same:
1. On a proof sheet, determine how many designs are to be included in each row.

2. Measure total number of vertical and horizontal inches available for design placement.

3. Placement will be measured from center of one design to center of next design. Spacing on the top and bottom edge of each vertical row will also be the same, although it may be different than spacing between patterns.

4. The spacing between vertical rows is determined by the size of the pattern and does not need to be the same as the vertical spacing in each row.

If All Motifs Are Not the Same:
1. Stencil several of each motif on practice paper and cut them out.

2. Determine the number of rows and number of motifs to be included in each row.

3. For vertical spacing of motifs when they are not the same, measure from bottom of one design to top of next design (between motifs).

4. Spacing between motifs needs to be the same. Spacing above and below each vertical row of motifs needs to be the same, but not necessarily the same as the internal spacing.

Measuring for Overall Grid Design:
1. Mark width of outside border around perimeter of floorcloth.

2. Locate center point of floorcloth and mark lightly with a chalk pencil; continue marking to edges of floorcloth.

3. Check measurements to corners and be certain floorcloth is square.

4. Refer to Figure 5 on page 19. From center point, measure to edge in equal increments (4", 8", 12", etc.).

Figure 5

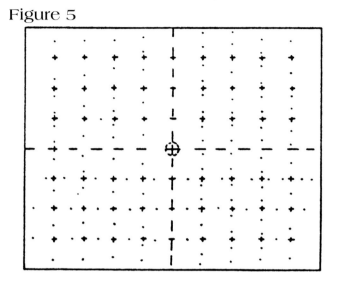

5. Connect lines vertically, horizontally, or diagonally.

6. Position motifs either on connecting lines or fitted within squares formed by connecting lines.

STRIPING

The addition of color striping as a border or to define an inset panel can add considerable depth, containment, and cohesiveness to a floorcloth design. There are several techniques and tools that can be used to apply these color stripes—a specialized lining brush, striping pens (from architectural drafting suppliers), striping tapes (originally used in the automotive industry for pinstriping), painter's tape, and finally, the simplest form, masking tape applied to a lightly marked line.

The major problem which results from using masking tape is the possibility of the strong adhesive backing removing the base-coat paint from the surface when the tape is removed. The following striping tips will minimize some of the problems which might occur:

• Allow base coat to cure three to four days before adding striping, if using masking tape.

• If tape appears to be very sticky, remove some adhesive by tapping strip of tape on a terry towel before attaching it to painted surface.

• Apply tapes in strips of 18" to 24" long. Press tape to surface with pressure against edge next to area to be painted. Paint this section, then remove tape within two to three minutes after painting. Do not allow tape to remain on surface for a long period of time. The longer it is attached, the more the adhesive will dry and the paint on the edge will seal tape to the surface.

• To avoid paint bleeding under tape, seal edge of tape, using the bowl of a teaspoon.

• If surface to be striped is slightly textured, paint underbleed can result, even if tape is carefully applied. Two techniques for applying paint and tape might be helpful in this case:

1. Before applying tape, apply a thin line of matte water-based varnish to edge of stripe area. This seal line will allow blotched lines to be wiped off after tape is removed. Allow varnish to dry thoroughly before taping.

2. Apply paint with a light brush stroke, going in same direction as tape and not brushing up to edge of tape from a perpendicular direction.

• If striping edge is not as crisp as you would like it to be because of slight underbleed in texture of surface, use a straightedge and a permanent marking pen and finish striping with a hard straight line. By selecting a marker that is slightly darker than the striping paint, the result will be a dimensional shadow to the stripe. Refer to Step 1 on page 20.

• Another option for correcting or sharpening a messy edge is to use an architectural ruling pen. These pens are available with various tip widths. They are loaded with paint, then used with a straightedge.

Striping

1 If edge of border is not straight because of an underbleed, it can be lined with a marking pen to make a clean line.

2 To paint an outside border, measure and mark stripe line with a chalk pencil.

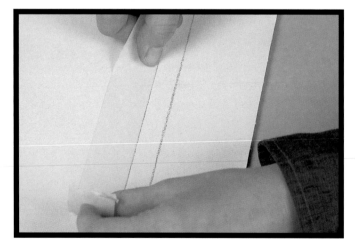

2a Attach masking tape inside marked line for border.

2b Remove tape after area has been painted to reveal outside border.

Striping

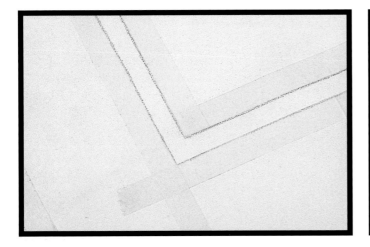

3 Interior Striping—Tape off area for striping, placing tape on both sides of area to be painted.

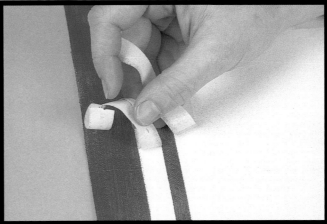

3a Remove tape to reveal painted stripe.

4 Positive/Negative Striping—Tape has been removed to reveal sponging, outside border, and interior striping.

5 Negative Striping—The tape will remain while other areas are painted. The taped area will be the background color.

Striping on an Outside Edge:

1. If hemming floorcloth, measure and mark width of hem with a chalk pencil. Refer to Steps 2 and 2a on page 20.

2. Measure and mark width of stripe with a chalk pencil.

3. Attach tape to inside edge of marked line.

4. Paint with a lightly loaded sponge brush or flat artist's brush.

5. Remove tape as soon as the 18" to 24" segment of striping has been painted. Refer to Step 2b on page 20. Add tape to next segment and continue in same manner.

Striping within Interior Sections:

There are two methods for measuring "interior stripes". Refer to Steps 3 and 3a on page 21.

Method #1

1. Measure to outside edge of stripe from outer edge of canvas. Mark.

2. Attach tape to outside edge of mark.

3. Attach another piece of tape exactly next to first strip toward inside of floorcloth. This piece represents the width of the stripe to be painted.

4. Attach another piece of tape exactly next to second strip. This will automatically be along inside edge of striping area.

5. Remove middle strip of tape (which was over striping area). Press down inside edges of other two pieces of tape.

6. Apply paint (which will adhere to floorcloth only between two tapes). Gently remove strips of tape.

Method #2

You may also measure for internal striping by measuring and spacing with a see-through quilter's ruler, marking edges of stripe. Attach tape outside these lines, along lines.

Positive & Negative Striping:

Use either positive or negative striping to section off and divide the design elements of the floorcloth. Refer to Step 4 on page 21.

The exterior border stripe can be any width to balance the edge. Interior striping should be used to accent. The most common widths for interior striping are ¼", ½", ⅝", and ¾" (seldom wider than the latter) on floorcloth sizes smaller than 6' x 9'.

If background area is sponged, a striping of negative space can add an interesting dimension to the surface. Refer to Step 5 on page 21.

For negative striping, the tape is not removed from the area until AFTER the sponging or the striping in each side is completed. This results in the negative stripe, which remains the same as the background color.

DESIGN BORDERS

A border gives a floorcloth a finished look. It is much like putting a frame around a photo or a painting. The border can be a solid painted border, a marbleized border, a continuing motif, or several rows of stripes. The border width will be determined by the overall finished size of the floorcloth or the look desired.

STENCILED BORDERS

Many precut stencils and stamps designed for wall borders make excellent borders around the edge of a floorcloth. Use a border around the edge in conjunction with a central motif or a sponged, marbleized, or solidly painted center. Measure in from outside edge of floorcloth or border stripe and make marks to keep stenciled border straight.

DETERMINING STYLE OF CORNERS

The style of the corners will depend on the flow of the design. Take a few minutes to look at the corners on floorcloths shown throughout this book to see how elements of the design dictate the best style of the corner. Some of the ways are:

• Special motif at corners as used on the Look-alike Marble floorcloth on page 32. Notice the large marbleized squares with the Greek Key stencil design. Refer to Figure 6.

Figure 6

Greek Key

• Butted corners can give an illusion of wider or longer rugs. Refer to Figure 7. An example of butted corners would be the Fruit Medley floorcloth on page 66.

Figure 7

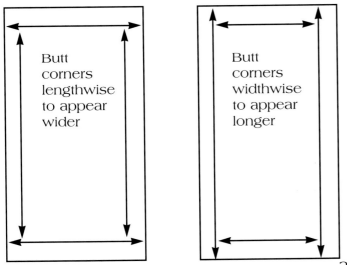

Butt corners lengthwise to appear wider

Butt corners widthwise to appear longer

• To make a miter corner, use a recipe card or a straightedge to mark and stencil up to the 45° angle. Refer to Figure 8.

Figure 8

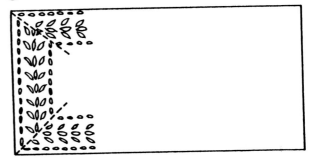

• Continue flow of design around corners. Refer to Figure 9. This is ideal for a vining-type design. Plan so design flow on all corners is as similar as possible. For example, you would not want one corner to flow "out and turn the corner" while another flows "in and turns the corner." This method is demonstrated on the Victoria's Rose Garden floorcloth on page 43.

Figure 9

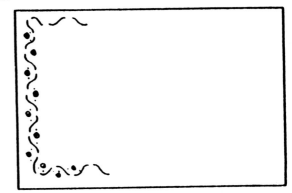

Background Treatments

SPONGING, RAGGING & MOPPING SUPPLIES

Colored Glazes:

Colored glazes are water-based, nontoxic mediums which can be added to neutral glazes in order to achieve a wide array of translucent tones. They come in a variety of premixed colors, which can be used alone, or mixed to achieve a special color. Colored glazes are excellent for sponging, ragging, or mopping, and cleanup is easy with soap and water.

Neutral Glazes:

Neutral glazes are water-based, nontoxic, and slightly transparent formulas that are perfect for ragging, sponging, or mopping techniques. They can be mixed with a variety of acrylic paint types, including acrylic craft paint, latex wall paint, and colored glaze. Follow manufacturer's instructions for ratios for mixing. Premixed sponging paints can also be used.

Sponging, Ragging, or Mopping Mitts:

Sponging, ragging, or mopping mitts are decorating tools with sponges, loops of fabric, or mop strings attached to create interesting textures with paint. They are extremely convenient to use and prevent messy hands. The mitts can be found at hardware or craft stores.

In place of mitts, a sea sponge or cellulose sponge can be used for sponging techniques; wadded rags for ragging techniques; and an actual string mop for mopping techniques.

Brushes:

French brushes, stippler brushes, and chamois tools can also be used to create textured effects. Each type of brush or tool gives a particular effect. Experiment with the techniques before trying it on the floorcloth.

Glazing the Background

1 Mix neutral wall glaze and decorator glaze colors according to individual project instructions to achieve desired tint.

2 Dip face of mitt in water. Wring out. Pat on towel. Pour small amount of glaze mixture on disposable plate. Place mitt on hand. Press face of mitt in glaze mixture. Pounce mitt on scrap of cardboard to blot.

3 Pat floorcloth randomly and repeatedly with face of mitt, overlapping to achieve consistent coverage. Change hand position frequently. Reload and rinse mitt as needed.

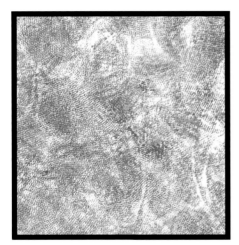

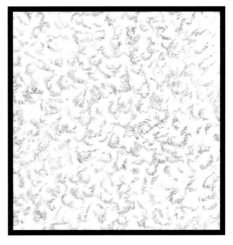

Ragged Finish

Sponged Finish

Mopped Finish

Hemming a Canvas Floorcloth

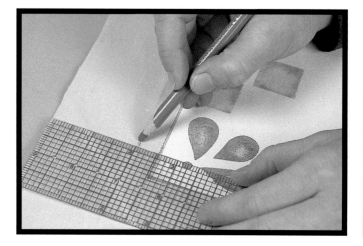

1 Mark hem on front of floorcloth with a chalk pencil.

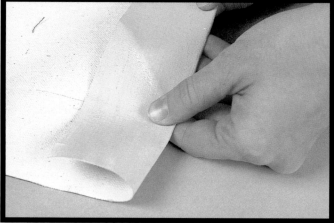

2 Turn hem under and finger-press.

3 Using scissors, cut off corner for a miter.

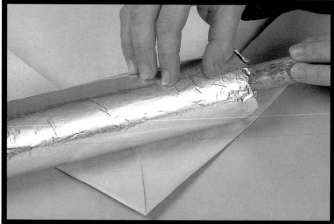

4 After applying glue, use a rolling pin covered with foil to smooth down hem.

HEMMING A CANVAS FLOORCLOTH

Hemming is done after stenciling or decorative painting is completed, and before finish is applied.

Preparing the Hem:
1. Measure and mark hem with a light-colored chalk pencil on front side of canvas. For smaller floorcloths, allow for a 1" hem. For larger pieces, allow for a 2" to 2½" hem. Use a see-through quilter's ruler that has a grid for 1" and 2" widths to mark hem. If cloth will be base-painted, mark canvas after base coat is applied.

2. Use a carpenter's square to assure finished piece is perfectly square after painting. Some slight adjustments may need to be made; use chalk pencil to correct hem edge and create hem lines. Use an artist's eraser to remove markings which have been changed. If a frame is used, square canvas after painting, and upon removing canvas from frame.

3. Fold edges under on hem lines and finger-press to crease.

4. Fold corners over and miter as follows:

• Crease each side, folding all the way out to edge of floorcloth.

• Fold corner up and realign creases with adjacent sides. Press this triangular corner piece.

• Using heavy scissors, cut off triangular piece. The corner piece will fold perfectly with no bulk to cause canvas piece to curl up on edges.

Gluing the Hem:
Select an adhesive that is fast-drying, such as a liquid cement. The slow-drying adhesives can cause rippling.

1. Flip floorcloth right side down on a flat, smooth surface. (For a larger floorcloth, put a drop cloth or large piece of plastic on garage floor or driveway to provide a surface.)

2. Using a stiff-bristled brush, apply adhesive to one side (or one section) at a time, working adhesive into canvas.

3. Using a rolling pin covered with aluminum foil or plastic wrap, roll hem one side at a time, assuring that all air pockets are removed.

4. To dry, turn floorcloth right side up and either clamp to a firring strip frame or apply weights to outer perimeter. (For larger cloths, when you do not have a frame, weight edges with bricks put inside plastic bags so they do not mark top of floorcloth.) Allow floorcloth to dry at least 24 hours before applying finish coats.

VARNISHING A FLOORCLOTH

Traditionally, the only varnishes recommended for floorcloths were oil-based because of durability and flexibility. The major disadvantage of oil-based products was they would "amber" or discolor, so it was impossible to have a finished floorcloth with a white-white background. In the recent past, a new generation of polyurethane has appeared that is water-based. They offer durability and flexibility, yet will not discolor the background of floorcloths. These products are environmentally "friendly" and much easier to work with and clean up.

Many floorcloth artists varnish the back of a canvas floorcloth and, although it is not necessary to coat the back of canvas, finishing the back gives a more complete look while adding weight to the canvas, which will help it lie flatter. It is recommended that at least one coat of varnish be applied to the back of canvas floorcloths to give a uniform finish and to seal the hemmed area (or raw edges, if not hemmed). Vinyl floorcloths require no varnish on the back.

FINISHING A FLOORCLOTH

Finishing Techniques:
1. Allow three to five days drying time after stenciling and hemming before application of finish coat. (Remember to "test the temperature" for dryness. If floorcloth feels cool, then it is not completely dry.)

2. Mist floorcloth with matte acrylic sealer spray. Matte sealer will assure that stencil paint will not bleed when varnish is applied. This is especially important if you cannot wait for paints to totally cure. Rushing the finish coat could cause paints to bleed.

3. Wipe off surface with tack cloth to remove dust that may have accumulated. Remove marks or smudges with an artist's eraser.

4. Varnish floorcloth in an area that is as dirt-free as possible.

5. If working on a large floorcloth, wear a pair of clean white socks and knee pads so surface remains as smudge-free as possible.

6. If using a roller, select a low-nap pad. Air bubbles will initially appear. Some will immediately pop. Apply first coat lightly. It is better to work varnish as little as possible when applying the first coat so as not to disturb painting.

7. Allow first coat to dry thoroughly. Do not rush drying time between application of coats. If working in a very humid climate, it may take several days between coats.

8. Lightly sand surface with a brown paper bag after first coat.

9. When applying second coat of varnish, if bubbles still appear, smooth surface with a 2" or 3" sponge brush that has been moistened in water. Do not use a "slip-slap" stroke with the brush; rather, pull brush smoothly in one direction across the surface.

10. Apply four to six coats of varnish to floorcloth. Before final coat, lightly sand with extra-fine-grit wet/dry sandpaper or #0000 steel wool. The final result will be an ultrasmooth finish.

11. It is recommended that a person making their first floorcloth should practice on a small surface area, such as a 2' x 3' piece. Finishing a floorcloth can be very tedious, and it is better to practice on a smaller surface area. Regardless of surface size, the finish coats are very important. The finest painting can be made to look second class if the time to apply the proper finish is not taken.

12. If a floorcloth begins to show wear, it can be recoated at any time. Wipe surface area to remove dirt and smudges. Apply one to two additional coats of varnish. Allow at least ten days drying time after recoating before returning the floorcloth to "active duty."

USING BACKINGS

There are backings that can be applied to a canvas floorcloth used on an area where sliding could occur, or to minimize the dangers that are not too different from hazards created by any throw rug. Consider these backing options:

• Use a thin jute or rubber padding (the type used under area rugs). Cut padding ⅛" to ¼" smaller on each side than the finished dimensions of the floorcloth.

• Apply an even coating of latex rubber backing to the back of the floorcloth, following manufacturer's instructions.

• The least desirable method, but still an option, is to apply double-stick carpet tape around the entire perimeter of backside of floorcloth.

Faux-stone
Floorcloths

In this chapter, you will learn how to bring the elegance of marble and the warmth of brick into your home. When applied to floorcloths, these textures look like the real thing.

Faux Brick
Instructions begin
on page 31

Faux Brick

Pictured on page 30

Designed by
Susan Goans Driggers

We love bricks outside. Now we can have the look of bricks indoors with a faux-brick finish. It is a snap to create this realistic faux-finish with grout tape and a few faux-finishing tools.

Floorcloth size: 30" x 43¼"

GATHER THESE SUPPLIES

Painting Surface:
Canvas or vinyl remnant,
    measuring 30" x 43¼"

Paints & Coatings:
Indoor/outdoor satin acrylics:
    Black
    Mojave Sunset
    Oxblood
    Pediment

Tools:
Chamois tool
French brush
Grout tape (or ¼"-wide masking tape)

Pencil
Sponge brush: 2"
Tape measure

Other Supplies:
Damp cloth rag

INSTRUCTIONS

Prepare:
NOTE: Lay floorcloth on a flat protected surface. Tape down edges with duct tape to keep in place while painting.

1. Refer to Preparing a Floorcloth on page 15. Measure, cut, and prepare floorcloth.

2. Using sponge brush, basecoat floorcloth with Pediment. Let dry.

Paint the Design:
1. Mix a darker gray color by combining one part Black to three parts Pediment.

2. Using chamois tool, stipple background slightly with this darker color.

3. Using pencil and tape measure, mark borders and bricks. Refer to photo on page 30 for placement. Corners are 6½" square; inside border is 2¼" wide; bricks are 3" x 6¼".

4. Using grout tape, mask ¼" stripe around inside of corners and inside border. Mix two parts Pediment to two parts Black to create a darker gray color than mixed previously. Using French brush, stipple stripe areas with this mix. Remove tape. Let dry.

5. Refer to Step 1. Tape off "bricks."

6. Refer to Step 2. Apply grout tape between "bricks."

7. Refer to Step 3. Stipple bricks, first with Mojave Sunset, then with Oxblood.

8. Using damp cloth rag, pounce colors to soften texture and contrast between colors. Remove tape. Let dry.

Finish:
1. Allow floorcloth to cure 72 hours before using. With indoor/outdoor satin acrylics, there is no need to seal floorcloth.✳

Step 1

Step 2

Step 3

Look-alike Marble
Instructions begin
on page 33

Look-alike Marble

Pictured on page 32

Designed by
Jane Gauss

With a few simple tools and some paint, you can create this "look-alike" marble inlaid floor inspired by a picture from a floor tile catalog.

Floorcloth size: 5' x 9'

GATHER THESE SUPPLIES

Painting Surface:
Canvas or vinyl remnant, measuring 5' x 9'

Paints & Coatings:
Acrylic craft paints:
  Dapple Gray
  Dove Gray
  Licorice
  Light Gray
  Wicker White
  Winter White
Matte acrylic sealer
Paint thickener
Water-based paint extender

Tools:
Carpenter's square
Chalk pencil
Marbleizing feather
Scissors
Sea or cellulose sponge
Sponge brushes: 2" (2)
Sponging mitt
Tape measure

Other Supplies:
Disposable plates
Lightweight cardboard
Masking tape: ⅝"-; ¼"-wide
Mixing stick
Single-overlay stencil:
  Greek Key

INSTRUCTIONS

Prepare:
1. Refer to Preparing a Floor-cloth on page 15. Measure, cut, and prepare floorcloth. There is no need to base-coat floorcloth since prime coat is white.

2. Using chalk pencil and tape measure, mark 4½" in from outside edge for outer border.

3. Refer to Figure 10 on page 34. Using scissors, cut template from lightweight cardboard measuring 8" x 8" square (or other desired size of tile). Using chalk pencil, draw vertical and horizontal alignment lines on template.

4. Measure and mark exact vertical and horizontal center of floorcloth. Using carpenter's square for 90° angles, extend center markings to outside edges. From center point, measure out 5½" in each direction of center.

5. Position center of template on center marking of floorcloth, turning square diagonally for a diamond shape. Using chalk pencil, mark around outside edge.

6. Refer to Figure 11 on page 34. Continue to draw 8" squares along center lines, aligning template with center lines and butting edge to previous line. Avoid space between squares. Masking tape will separate marking lines to create grout spaces.

7. Using ⅝" masking tape, mask around outside edge of squares to "contain" marked-off squares.

8. After marking and masking off horizontal and vertical lines of squares, center ¼" masking tape on each diagonal line.

Paint the Design:
1. Using sponge brushes, base-coat every other square with Licorice and remaining squares with Light Gray. Do not remove tape yet.

Marbleizing:
1. For black squares, place puddles of the following paints on disposable plate: Dapple Gray, Licorice, and very little Light Gray. Squeeze small amounts of paint thickener and paint extender over these paints. Using mixing stick, swirl paints together. Do not completely mix.

2. Dip dampened sea sponge into paint. Pounce sponge on disposable plate to distribute colors into sponge. Then pounce or dab sponge onto black squares. Refill sponge with paint as necessary. Sponge up to and over tape, but block-off sponging so colors do not spill into adjacent squares.

3. For gray squares, place puddles of the following paints on disposable plate: Dove Gray, Wicker White, and Winter White. Sponge with these colors, repeating sponging technique as described above. Let dry.

Veining:

1. Vein black squares with marbleizing feather. Place puddles of Light Gray and Wicker White on disposable plate. Squeeze small amount of paint extender and thickener over these paints. Using mixing stick, swirl paints together. Do not completely mix.

2. Pull feather through paint and lightly pull veins onto squares. (Practice veining first, before veining on floorcloth.)

3. To keep veins muted, lightly sponge over portions of veins with sponge. The entire effect of floorcloth can be ruined if veins look like lightning bolts.

Striping:

1. Remove masking tape.

2. Measure outside edge of marble squares area. Mask around outside edge of squares. One row of tape will be on top of marbled squares, the other will be placed ⅝" away from the first, so a ⅝" space is between them.

3. Measure in ⅝" from outside edge and mark with masking tape.

4. Paint outside and inside stripes with Licorice.

Stenciling:

1. Stencil Greek Key design inside Licorice outer border. Stencil with Licorice. Refer to photo on page 32 for placement. Let dry.

Finish:

1. Refer to Varnishing a Floorcloth on page 27. Seal floorcloth with matte acrylic sealer.✳

Figure 10

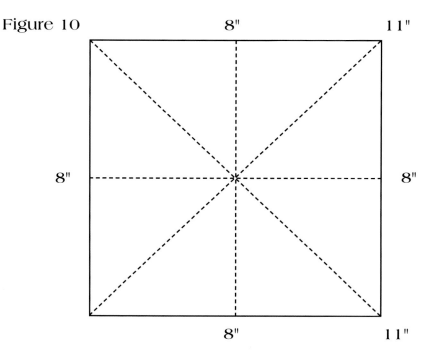

Figure 11

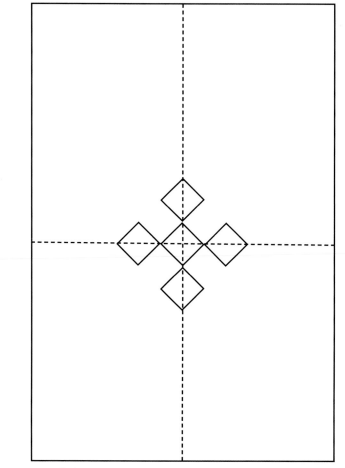

34

Stenciled Floorcloths

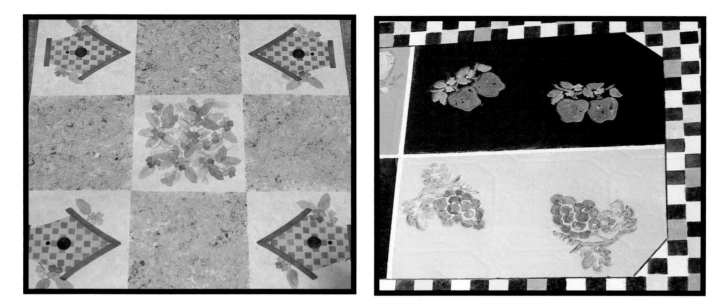

Using stenciling to decorate your floorcloths gives you limitless possibilities. There are stencils of almost every size and design style. You will find simple stencils with just a single overlay as well as very complex stencils with multiple overlays. In this chapter, you will learn to use a variety of stencil types and stencil paints to create beautiful floorcloths. There are projects for children, designs to brighten up your kitchen, and floorcloths to make elegant statements in your dining or living room.

Stenciling

STENCILING SUPPLIES

Stencils are pieces of flexible, paint-resistant material, which have cut-out areas forming a pattern. They can be used to decorate almost any surface and are available in a wide variety of patterns as well as application styles.

Geometric Background Stencils:

Geometric background stencils are reusable stencils that act as a grid for quickly and easily establishing a background pattern for floorcloths. This background pattern can stand alone, or be the basis for adding other decorative elements with stamps, blocks, or other stencils.

Multi-overlay Stencils:

Multi-overlay stencils are precut with one overlay per color. Design registration marks on each sheet make it easy to align overlays for professional-looking results. Some multi-overlay stencils have coordinating spot motif stencils that are perfectly sized for stenciling decorative accessories.

Single-overlay Stencils:

Single-overlay stencils are precut stencils that do not have overlays. They can be used on a number of surfaces. Some single-overlay stencils feature uniquely shaped edges to be used as design elements.

Stencil Brushes:

Stencil brushes are available in a variety of sizes. The size of brush you use depends on the size of the openings in your stencil. To stencil a small, delicate print, you may choose a ¼" brush; stenciling a large design might require a 1" brush. You need to have a separate brush for each color you plan to stencil in one day. To achieve quality stencil prints, you must allow your brush to dry thoroughly after cleaning, before you use it again.

Stencil Rollers:

Stencil rollers are ideal for achieving quick prints and covering background areas. The roller also protects delicate areas of the stencil and works well in areas of the stencil where using a brush could cause the stencil to move. Rollers are made of sponge and work exceptionally well with stencil gels.

Brush Cleaner:

Brush cleaner is a liquid gel used for cleaning and conditioning brushes. It is necessary to clean brushes properly for best results when using them again.

Acrylic Craft Paints:

Acrylic craft paints offer quick drying time and a wide variety of premixed colors. With acrylics, it is also easy to mix custom colors to coordinate with home decor. On most surfaces, mistakes can be wiped off before paint dries.

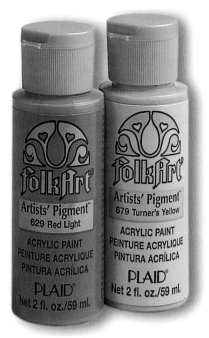

Dry Brush Stencil Paints:

Dry brush stencil paints combine quality and convenience. Their creamy, no-drip formulation makes them easy to use on many surfaces. They are available in a wide array of decorator colors and come in easy to hold palm-sized jars.

Gel Paints:

Gel paints are paints formulated for stenciling. They produce a translucent, watercolor look. The thick formulation holds well on a brush or roller, and can be blended, toned, and shaded with ease.

Other Supplies:

• Blender gel—for allowing paint to stay wet longer without changing color or consistency. It is recommended when stenciling with a roller. The blender serves as a base for the color and helps keep paint from drying out.

• Chalk pencils—for marking floorcloth surface.

• Disposable plates or palette—for holding and mixing paint.

• Paper towels—for wiping brushes.

• Stencil adhesive—for securing stencil to stenciling surface.

• Stencil tape—for securing stencil to floorcloth.

• Tape measure or ruler—for measuring.

Using Gel Paints & Acrylic Craft Paints

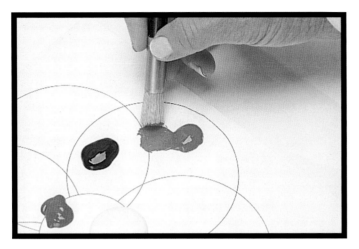

1 Squeeze out dime-sized amounts of gel or acrylic paints on a disposable plate or palette. Hold brush perpendicular to the palette and pull out a small amount of paint. Twist brush to concentrate paint in center of brush.

2 Remove excess paint by lightly pouncing loaded brush on a paper towel.

3 Bring paint into cut-out area of stencil with a light pouncing stroke or circular stroke, keeping brush perpendicular to surface. Use more pressure to shade edges of cut-out area or to create an opaque print.

Stenciled Noah's Ark

Using a Roller

1 Prime a dry roller by misting with water; then towel dry. Squeeze about one teaspoon of blender gel onto a disposable plate or piece of aluminum foil. Roll the roller in blender.

2 Squeeze gel paint or acrylic paint onto a disposable plate or piece of aluminum foil. Roll the primed roller through the paint. Distribute paint throughout roller so it is evenly covered. Blot roller to remove excess paint by rolling on paper towels.

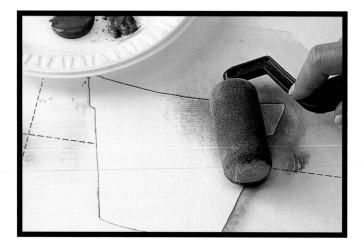

3 Roll back and forth over stencil, beginning with a light stroke to test amount of paint.

4 The finished stencil.

Using Dry Brush Stencil Paints

The creamy formulation of dry brush stencil paints makes them easy to use. Because this paint dries more slowly than a liquid paint, you should wait at least one to two hours before stenciling overlays to avoid smudging. At this point, the paint will still not be dry so you will need to be careful when placing overlays. Allow paint to cure for three to five days before applying sealer or varnish.

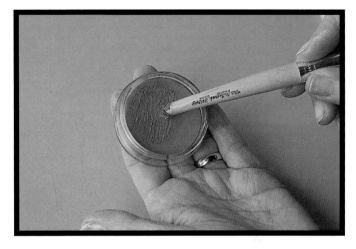

1 Gently remove seal from paint with the handle end of a brush. The seal protects the paint from drying out; it is necessary to remove the seal after each period of inactivity.

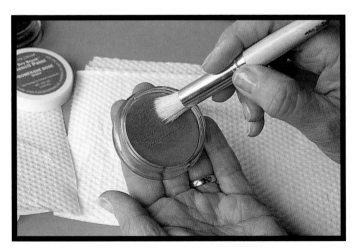

2 Swirl brush in paint to load.

3 Swirl brush on an uncut portion of stencil to disperse paint evenly through bristles. Do this each time you load the brush. Begin stenciling by gently swirling paint into cut-out areas.

Using Multi-overlay Stencils

1 Stencils with more than one overlay have registration marks to aid in lining up properly. One overlay is used for one color; the next overlay is used for another color. Use the overlay marked "A" to stencil the first color.

2 Line up the registration marks and stencil the second color with the overlay marked "B".

3 Line up the registration marks and stencil the third color with the overlay marked "C".

4 The finished stencil.

Victoria's Rose Garden
Instructions begin on page 44

Victoria's Rose Garden

Pictured on page 43

Designed by
Jane Gauss

Have blooming roses year-round with this beautiful floorcloth. The stenciled lace centerpiece is a wonderfully subtle complement to the colorful flowers.

Floorcloth size: 2' x 3'

GATHER THESE SUPPLIES

Painting Surface:
Canvas or vinyl remnant,
    measuring 2' x 3'

Paints & Coatings:
Acrylic craft paints:
    Bright Peach
    Clay Bisque
    Green Meadow
Dry brush stencil paints:
    Cameo Peach
    Promenade Rose
    Wild Ivy Green
Matte acrylic sealer
Repositionable spray adhesive

Tools:
Chalk pencil
Paint roller
Ruler
Sponge brush: 1"
Stencil brushes: ⅛"; ¾"
Tape measure

Other Supplies:
Masking tape: ¼"-wide
Polyester lace: ½ yard
Single-overlay stencil:
    Victoria's Rose

INSTRUCTIONS

Prepare:
1. Refer to Preparing a Floorcloth on page 15. Measure, cut, and prepare floorcloth.

2. Using paint roller, base-coat floorcloth with Clay Bisque. Let dry.

3. Refer to Victoria's Rose Garden Diagram on page 45. Using chalk pencil and tape measure, mark ½" for outside border.

Paint the Design:
1. Mask off inside of chalk line with ¼"-wide masking tape. Using sponge brush, paint outside border with Bright Peach. Let dry.

2. Position lace in center of floorcloth. On 2' x 3' piece, the lace is positioned 6⅝" in from edges of floorcloth.

3. Locate exact center of floorcloth and place center of lace piece on this point. Hold lace in place with small strips of masking tape.

4. Lightly apply repositionable spray adhesive onto back of lace by lifting one half of lace, spraying back, then placing lace back onto cloth; repeat with other half of lace. This way position of lace has not changed.

5. Using ¾" stencil brush, stencil through lace with Bright Peach. Remove lace panel. Let dry.

6. Measure and mark floorcloth for green inner striping around lace panel and also inside outer peach border striping. Mask along markings. This will leave a ⅛"-wide strip. Using ⅛" stencil brush, paint with Green Meadow. Use a vertical back-and-forth stroke within masked areas to create narrow stripes.

7. Using Victoria's Rose stencil and stencil brushes, stencil rose design in the 6"-wide area between narrow green stripes. Refer to photo on page 43.

8. Using ¾" stencil brush, stencil a rose in each corner and roses in "mirror images" on sides with Cameo Peach. Shade with Promenade Rose. If uncertain about placement, stencil several prints on practice paper, cut them out, and lay them on floorcloth. To create mirror image, stencil on the "right" and then the "wrong" side of the stencil. Wipe stencil before flipping stencil sheet.

9. Using ⅛" stencil brush, stencil leaves on the ends of floorcloth with Wild Ivy Green.

Finish:
1. Refer to Varnishing a Floorcloth on page 27. Seal floorcloth with matte acrylic sealer.✳

Victoria's Rose Garden Diagram

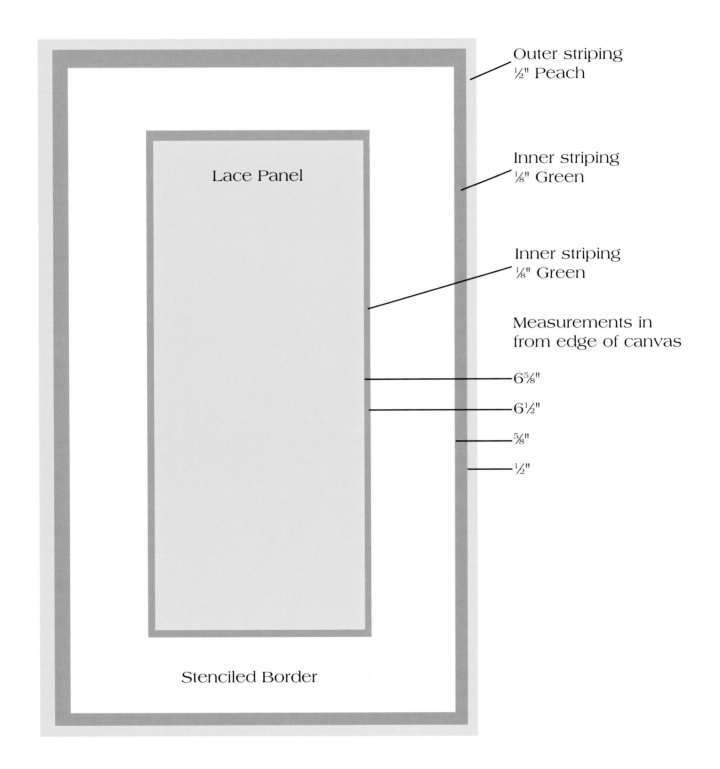

Outer striping
½" Peach

Inner striping
⅛" Green

Inner striping
⅛" Green

Measurements in
from edge of canvas

6⅝"

6½"

⅝"

½"

Lace Panel

Stenciled Border

Intertwining Ivy
Instructions begin
on page 47

Intertwining Ivy

Pictured on page 46

Designed by
Jane Gauss

Once you step on this lovely floorcloth covered in ivy, you will not know whether you are indoors or out. It would look especially lovely in a sunroom.

Floorcloth size: 2' x 3'

GATHER THESE SUPPLIES

Painting Surface:
Canvas or vinyl remnant,
  measuring 2' x 3'

Paints & Coatings:
Acrylic craft paints:
  Mint Green
  Sap Green
  Warm White
Dry brush stencil paints:
  Sherwood Forest Green
  Truffles Brown
  Wild Ivy Green
Faux-finishing glaze: Neutral
Matte acrylic sealer

Tools:
Chalk pencil
Paint roller
Sponging mitt or sea sponge
Stencil brushes: 1" (2)
Straightedge
Tape measure

Other Supplies:
Disposable plates
Marking pen: soft gray, ⅛"-wide
  tip
Masking tape: ¼"-; ½"-wide
Single-overlay stencil:
  Gardener's Ivy

INSTRUCTIONS

Prepare:
1. Refer to Preparing a Floor-cloth on page 15. Measure, cut, and prepare floorcloth. There is no need to base-coat floorcloth since prime coat is white.

2. Refer to Striping on pages 19–22. Create positive and negative striping. The outside green stripe is ¾" wide. The inside white stripe is ¼" wide.

3. Using chalk pencil and tape measure, mark in ¾" on inner edge of green border. Mask off along mark.

Paint the Design:
1. Using paint roller, paint outside edge (outside of tape) with Sap Green. Leave tape in place for sponging inside area.

2. Using disposable plate, mix Neutral glaze with Mint Green. Refer to Glazing the Background on page 25. Using sponging mitt, sponge inside rectangle (inside of tape).

3. Mix Neutral glaze with Warm White and lightly sponge over Mint Green. Let dry.

4. Using Gardener's Ivy stencil and 1" stencil brush for each color, stencil on sponged area. Use a repeat section of pattern to connect ends and center section. Stencil leaves with a mix of Wild Ivy Green and Sherwood Forest Green. Stencil inside leaves and stems with a mix of Sherwood Forest Green and Truffles Brown.

5. Remove tape. The section under tape will remain white (this is the "negative" stripe). Using straightedge and marking pen, make a ⅛"-wide light gray line, along sponged edge. This will give a shadowed look to the striping. Let dry.

Finish:
1. Refer to Varnishing a Floorcloth on page 27. Seal floorcloth with matte acrylic sealer.✳

Ivy & Bird

Pictured on page 49

Designed by
Kathi Malarchuk

Using more than one stencil to create a floorcloth design can result in a beautiful, one-of-a-kind effect. This happy mix of ivy and birds creates a classic, understated look.

Floorcloth size: 4' x 6'

GATHER THESE SUPPLIES

Painting Surface:
Canvas or vinyl remnant,
    measuring 4' x 6'

Paints & Coatings:
Blender gel
Faux-finishing colored glazes:
    Ivy Green
    Mushroom
    Neutral
Matte acrylic sealer
Stencil gel paints:
    Cactus
    Daffodil Yellow
    Ivory Lace
    Napa Grape
    Twig
    Wild Ivy

Tools:
Chalk pencil
Craft knife
Ragging mitt
Ruler
Stencil brushes: 1" (6)
Tape measure

Other Supplies:
Disposable plate
Masking tape: ½"-wide
Multi-overlay stencils:
    Apple Blossom & Bird's Nest
    Ivy

INSTRUCTIONS

Prepare:
1. Refer to Preparing a Floorcloth on page 15. Measure, cut, and prepare floorcloth.

2. Refer to Glazing the Background on page 25. Using disposable plate, mix 2 ounces of Mushroom glaze with 12 ounces of Neutral glaze for a light tint.

3. Using ragging mitt, rag entire floorcloth with the glaze mixture. Let dry.

4. Using chalk pencil and tape measure, mark in 12" from each side. Mask off a 2" border. On shorter sides, measure in 6" from corners and mark. On longer sides, measure in 8" from corners and mark. Join marks with masked lines 2" apart to create slanted corners. Using craft knife, trim away excess being careful not to cut floorcloth.

5. Measure 4" from outer edges for outer border. Mask off.

6. Mix Ivy Green glaze with remaining Neutral glaze. Prepare ragging mitt and rag borders with glaze mixture. Let dry. Remove tape.

7. Place masking tape around inner edge of outer border. Using 1" stencil brush, pounce with Twig. Let dry. Remove tape.

Paint the Design:
1. Refer to Using Multi-overlay Stencils on page 42. Using Ivy stencil and stencil brushes, one for each color; stencil ivy vines in four corners and center.

2. Stencil leaves with Wild Ivy; veins and vines with Twig.

3. Stencil overlay B leaves with Cactus; veins and vines with Twig.

4. Stencil overlay C veins and vines with Twig. Let dry.

5. Using Apple Blossom & Bird's Nest stencil, position birds to perch on ivy vines.

6. Stencil overlay A small bird with Daffodil Yellow. Shade with Napa Grape. Stencil overlay B with Napa Grape mixed with blender gel. Stencil feet with Twig. Stencil overlay C with Napa Grape.

7. Stencil overlay A large bird with Ivory Lace mixed with Daffodil Yellow. Shade edges with Napa Grape. Stencil overlay B with Napa Grape mixed with blender gel. Stencil feet with Twig. Stencil overlay C with Napa Grape. Let dry.

Finish:
1. Refer to Varnishing a Floorcloth on page 27. Seal floorcloth with matte acrylic sealer.✳

Ivy & Bird
Instructions begin
on page 48

Up, Up & Away

Pictured on pages 52–53

Designed by
Susan Goans Driggers

Fluffy clouds and big balloons make this floorcloth fun in any child's room.

Floorcloth size: 2' x 3'

GATHER THESE SUPPLIES

Painting Surface:
Canvas or vinyl remnant,
    measuring 2' x 3'

Paints & Coatings:
Acrylic craft paints:
    Cobalt
    Raspberry Sherbet
    Tartan Green
Matte acrylic sealer

Tools:
Chalk pencil
Sponge brush: ½"
Stencil brush: 1"
Tape measure

Other Supplies:
Masking tape: ½"-wide
Single-overlay stencil:
    Up, Up & Away

INSTRUCTIONS

Prepare:
1. Refer to Preparing a Floor-cloth on page 15. Measure, cut, and prepare floorcloth. There is no need to base-coat floorcloth since prime coat is white.

2. Using chalk pencil and tape measure, mark ½" inside edges. Mask off outside of chalk line so outer ½" border is covered with tape.

Up, Up & Away Diagram

Paint the Design:
1. Refer to Up, Up & Away Diagram on page 51 for placement of stencil. Using Up, Up & Away design and stencil brush, stencil balloons and clouds, alternating patterns with Cobalt, Raspberry Sherbet, and Tartan Green.

2. Stencil all clouds with Cobalt. Be certain to stencil clouds lightly, keeping concentration of color to outer edges, so they appear airy and light. Notice some balloons have areas of light and dark concentrations. Let dry.

3. Remove masking tape from outer edge and replace on opposite side of chalk line so outer ½" of floorcloth is exposed.

4. Using ½" sponge brush, paint outer ½" border with Cobalt. Remove tape carefully. Let dry.

Finish:
1. Refer to Varnishing a Floorcloth on page 27. Seal floorcloth with matte acrylic sealer.✳

Up, Up & Away
Instructions begin
on page 51

Welcome Home
Instructions begin
on page 55

Welcome Home

Pictured on page 54

Designed by
Avis Everett

What better way to welcome family and friends than with this plush-looking floorcloth.

Floorcloth size: 36" half circle

GATHER THESE SUPPLIES

Painting Surface:
Canvas or vinyl remnant,
  measuring 36" half circle

Paints & Coatings:
Indoor/outdoor satin acrylics:
  Mustard
  Olive
  Vanilla

Tools:
Chalk pencil
Paint roller
Stencil brushes: ¼"; ⅝"; 1"
Tape measure

Other Supplies:
Masking tape: 2"-wide
Single-overlay stencils:
  Classic Ivy
  Country Greetings
Stencil tape

INSTRUCTIONS

Prepare:
1. Refer to Preparing a Floorcloth on page 15. Measure, cut, and prepare floorcloth.

2. Using paint roller, basecoat floorcloth with Vanilla. Let dry.

3. Using chalk pencil and tape measure, find center of floorcloth by measuring across top on straight edge, then down fattest part of center. Place Country Greetings stencil in center of floorcloth with "C" of "Welcome" on center point. Using stencil tape, secure stencil to floorcloth.

Paint the Design:
1. Using ⅝" stencil brush, stencil letters and leaves of pineapples with Olive. Remove stencil tape and clean stencil. Let dry.

2. Replace stencil on floorcloth. Mask off leaves on pineapples. Using 1" stencil brush, stencil pineapples with Mustard. Let dry.

3. Place top of Classic Ivy stencil even with straight edge of floorcloth. Using ¼" stencil brush, stencil ivy with Olive. Remember to tape stencil securely to floorcloth and to keep paint very dry on brush.

4. Shade leaves by dabbing lighter in some areas (such as centers) and heavier toward outer edges. Continue stenciling ivy around curved part of floorcloth in same manner.

Finish:
1. Let floorcloth dry. With indoor/outdoor satin acrylics, there is no need to seal floorcloth.✳

Grapevine
Instructions begin
on page 57

Grapevine

Pictured on page 56

Designed by
Kathi Malarchuk

This floorcloth uses a wall border stencil to create a free-flowing design.

Floorcloth size: 4' circle

GATHER THESE SUPPLIES

Painting Surface:
Vinyl remnant or canvas,
    measuring 4' circle

Paints & Coatings:
Acrylic craft paint:
    Buttercrunch
Dry brush stencil paint:
    Wild Ivy Green
Faux-finishing glazes:
    Neutral
    Persimmon
    Russet Brown
Matte acrylic sealer

Tools:
Chalk pencil
Paint roller
Stencil brush: ½"
Stippler brush
Tape measure

Other Supplies:
Disposable plate
Masking tape: 1"-wide
Single-overlay stencil:
    Nature's Vineyard

INSTRUCTIONS

Prepare:
1. Refer to Preparing a Floorcloth on page 15. Measure, cut, and prepare floorcloth.

2. Using paint roller, base-coat floorcloth with Buttercrunch. Let dry.

Paint the Design:
1. Using disposable plate, mix equal parts of Persimmon and Russet Brown glazes. Blend half of glaze mixture with Neutral glaze to achieve a light tint. Reserve remaining half of dark glaze mixture.

2. Using a stippler brush, pounce entire floorcloth with light glaze mixture. Let dry.

3. Using a chalk pencil and tape measure, mask off a 4" border at edge of floorcloth.

4. Stipple border with reserved dark glaze mixture. Remove tape. Let dry.

5. Using Nature's Vineyard stencil and ½" stencil brush, stencil grapevines and grapes in lighter area of floorcloth with Wild Ivy Green. Refer to photo on page 56 for placement. Let cure for four days.

Finish:
1. Refer to Varnishing a Floorcloth on page 27. Seal floorcloth with matte acrylic sealer.✳

English Topiary

Pictured on page 58

Designed by
Kathi Malarchuk

Shades of green add warmth to any room, just as live plants do. This floorcloth gives the best of both worlds by combining ivy topiaries with a beautiful green background.

Floorcloth size: 28" x 42"

GATHER THESE SUPPLIES

Painting Surface:
Canvas or vinyl remnant,
    measuring 28" x 42"

Paints & Coatings:
Acrylic craft paint:
    Mystic Green
Dry brush stencil paints:
    Ol' Pioneer Red
    Terra Cotta
Faux-finishing glazes:
    Bark Brown
    Deep Woods Green
    Ivy Green
    Neutral
    White
Matte acrylic sealer

Tools:
Chalk pencil
Disappearing marker
Flat brush: ½"
Paint roller
Round brush: #3
Stencil brush: ½"
Tape measure

Other Supplies:
Disposable plate
Masking tape: ½"-wide
Plastic wrap
Printing block: Ivy
Single-overlay stencil:
    Pots & Planters

INSTRUCTIONS

Prepare:
1. Refer to Preparing a Floorcloth on page 15. Measure, cut, and prepare floorcloth.

2. Using paint roller, base-coat floorcloth with two coats of Mystic Green. Let dry between coats.

Continued on page 59

English Topiary
Instructions begin
on page 57

Continued from page 57

3. Using disposable plate, mix two parts White glaze with one part Neutral glaze.

4. Crumple a 12" square of plastic wrap. Dip wrap in glaze mixture. Pat glaze lightly over entire surface of floorcloth, creating a mottled look. Let dry.

5. Using chalk pencil and tape measure, mask off a 4" border on all sides.

6. Using ½" flat brush, brush Deep Woods Green glaze on border. Crumple a 12" square of plastic wrap. Pat crumpled plastic wrap over glaze, creating a mottled look. Remove tape. Let dry.

Paint the Design:
1. Using ½" stencil brush and Pots & Planters stencil, stencil three pots in center of floorcloth with Terra Cotta. Shade pots with Ol' Pioneer Red.

2. Using disappearing marker, draw heart shape for center topiary and circle shapes for left and right topiaries.

3. Refer to Block Printing on page 104. Using Ivy printing block, block-print leaves for topiaries with equal parts of Deep Woods Green and Ivy Green glazes.

4. Using #3 round brush, paint twining stems, vines, and tendrils with equal parts of Bark Brown and Neutral glazes. Let dry.

Finish:
1. Refer to Varnishing a Floorcloth on page 27. Seal floorcloth with matte acrylic sealer.✷

Birdhouse

Pictured on page 60

Designed by
Kathi Malarchuk

This floorcloth combines stenciling with stencil blocks to create a crisp, clean design that has softer details.

Floorcloth size: 40" x 60"

GATHER THESE SUPPLIES

Painting Surface:
Canvas or vinyl remnant, measuring 40" x 60"

Paints & Coatings:
Acrylic craft paint:
    Spring White
Dry brush stencil paints:
    Andiron Black
    Ecru Lace
    True Blue
Faux-finishing glazes:
    Alpine Green
    Blue Bell
    Neutral
    Persimmon
    Sunflower
Matte acrylic sealer

Tools:
Chalk pencil
Paint roller
Ragging mitt
Round brush: #3
Stencil brushes: ½" (3)
Tape measure

Other Supplies:
Disposable plate
Masking tape: ½"-wide
Paper towels
Printing block:
    Little Garden Flowers
Single-overlay stencils:
    Birdhouses
    Checkerboard

INSTRUCTIONS

Prepare:
1. Refer to Preparing a Floorcloth on page 15. Measure, cut, and prepare floorcloth.

2. Using paint roller, base-coat floorcloth with Spring White. Let dry.

3. Using disposable plate, mix 2 ounces Sunflower glaze with enough Neutral glaze to create a medium tint.

4. Dampen ragging mitt. Blot on paper towel to remove excess moisture. Dip ragging mitt in glaze mixture. Pat glaze lightly over entire surface of floorcloth, creating a ragged texture. Let dry.

5. Using chalk pencil and tape measure, mask off a 5"-wide border on all sides. Measure 1" from inner edge of border and mask off. Measure 2" from outer edge of floorcloth and mask off.

Paint the Design:
1. Mix Blue Bell glaze and enough Neutral glaze to create a deep tint. Dampen ragging mitt. Blot on paper towel to remove excess moisture. Dip ragging mitt in glaze mixture. Pat glaze on borders inside taped lines. Remove tape. Let dry.

2. Measure and mask off three rows of five squares, each 10" square.

3. Mix Neutral glaze with Alpine Green glaze. Dampen ragging mitt. Blot on paper towel to remove excess
Continued on page 61

Birdhouse
Instructions begin
on page 59

Continued from page 59

moisture. Dip ragging mitt in glaze mixture. Pat glaze on alternate squares. Refer to photo on page 60 for placement. Remove tape. Let dry.

4. Using stencil brush, stencil birdhouse walls with Ecru Lace.

5. Stencil roofs and bases with True Blue.

6. Using Checkerboard design, stencil checks on walls of birdhouses with True Blue.

7. Stencil entrance holes and perches with Andiron Black.

8. Using small flower petal from the Little Garden Flowers printing block, block small flowers in center square, around birdhouses, and on light band between the two blue ragged borders with Blue Bell.

9. Block-print leaves with Alpine Green.

10. Using round brush, add flower centers with Persimmon.

11. Paint stems and tendrils with Alpine Green. Let dry.

Finish:
1. Refer to Varnishing a Floorcloth on page 27. Seal floorcloth with matte acrylic sealer.✳

Hands & Ribbons

Pictured on page 62

Designed by
Jane Gauss

What a great project for rainy days when children can not go out and play. Enlist their hands and help when making this adorable floorcloth.

Floorcloth size: 4' x 4' octagon

GATHER THESE SUPPLIES

Painting Surface:
Canvas or vinyl remnant,
  measuring 4' x 4'

Paints & Coatings:
Acrylic craft paints:
  Cobalt
  Dioxazine Purple
  Medium Yellow
  Napthol Crimson
  Teal
Découpage finish
Matte acrylic sealer
Water-based paint extender

Tools:
Chalk pencil
Craft or utility knife
Heavy-duty scissors
Ruler
Sponge brushes: ½" (6)
Stencil brush: 1"

Other Supplies:
Hand soap
Moist sponge
Satin ribbon: ⅜"-wide, blue;
  green; purple; red; yellow
  (1 yd. each)
Single-overlay stencil:
  Color Crayons
White craft glue

INSTRUCTIONS

Prepare:
1. Refer to Preparing a Floorcloth on page 15. Measure, cut, and prepare floorcloth.

2. Using heavy-duty scissors, form octagon shape of floorcloth by angling corners. Using chalk pencil and tape measure, find center of each 4' side and mark. From center point, measure 9" from center mark to left and 9" to right of center.

3. Repeat on each side of square floorcloth. Connect marks with diagonal line from one side of floorcloth to adjacent side, to mark corners.

4. Cut off corners. There is no need to base-coat floorcloth unless background color other than white is desired.

5. Using white craft glue, adhere ribbons from outside row around edge of floorcloth to inside row. Refer to photo on page 62 for ribbon color placement. Begin approximately ¼" from outside edge.

6. Using craft knife, miter ribbon at each corner. Work one 18" to 24" section at a time, adhering first with white craft glue, checking spacing, and patting in place with moist sponge.

7. Begin next color approximately ⅛" in from first stripe.

8. Lay floorcloth flat until completely dry.
Continued on page 63

Hands & Ribbons
Instructions begin on
page 61

Continued from page 61

9. Using ½" sponge brush, apply two thin coats of découpage finish over entire ribbon border. Let dry.

Paint the Design:
1. Mix one teaspoon of paint with four to six drops of extender. Mix thoroughly.

2. Using sponge brush, paint palm of child's hand, being certain entire palm is covered, yet paint is not dripping off hand.

3. Gently press child's hand in place on floorcloth, then lift directly up from surface, not moving hand from side to side.

4. Thoroughly wash child's hand with soap and water and dry before repainting with another color for next print. Use a different sponge brush for each color. Make handprints in an overall pattern on entire floorcloth inside ribbon border, turning hands in various directions and using various paint colors.

Note: If several children are contributing their hands to the project, use a permanent, fine-tip marking pen and inscribe each print with child's name and age.

5. Using 1" stencil brush and Color Crayons stencil, stencil crayons randomly among handprints. Vary colors from crayon to crayon. Let dry.

Finish:
1. Refer to Varnishing a Floorcloth on page 27. Seal floorcloth with matte acrylic sealer.✳

Damask

Pictured on page 64

Designed by
Kathi Malarchuk

The bold colors and dramatic background design of this floorcloth are what make it special. Any room with this accent is sure to make a statement.

Floorcloth size: 4' x 6'

GATHER THESE SUPPLIES

Painting Surface:
Canvas or vinyl remnant, measuring 4' x 6'

Paints & Coatings:
Indoor/outdoor satin acrylics:
  Barn Red
  Fairway Green
  Khaki
  Vanilla

Tools:
Chalk pencil
Paint roller
Sponge brushes: ¼"; ¾"; 1" (3)
Stencil brush: 1"
Tape measure,

Other Supplies:
Geometric background stencil: Damask
Masking tape: ½"-wide

INSTRUCTIONS

Prepare:
1. Refer to Preparing a Floorcloth on page 15. Measure, cut, and prepare floorcloth.

2. Using paint roller, base-coat floorcloth with two coats of Vanilla. Let dry between coats.

3. Refer to Damask Pattern on page 65. Using chalk pencil and tape measure, mask off borders.

Paint the Design:
1. Start at outside edge and work inward. Using 1" sponge brush, paint outside border with Fairway Green.

2. Using another 1" sponge brush, paint 4" border with Barn Red.

3. Using another 1" sponge brush, paint 1" border with Khaki.

4. Using ¼" sponge brush, paint ¼" border with Fairway Green.

5. Using 1" sponge brush, paint 1" border with Khaki.

6. Using ¾" sponge brush, paint ¾" border with Fairway Green.

7. Measure and mark exact horizontal and vertical middle of center section of floorcloth.

8. Using Damask stencil and 1" stencil brush, stencil design with Barn Red in center of floorcloth.

Finish:
1. Let floorcloth cure for 72 hours. With indoor/outdoor satin acrylics, there is no need to seal floorcloth.✳

Damask
Instructions begin on
page 63

Damask Pattern

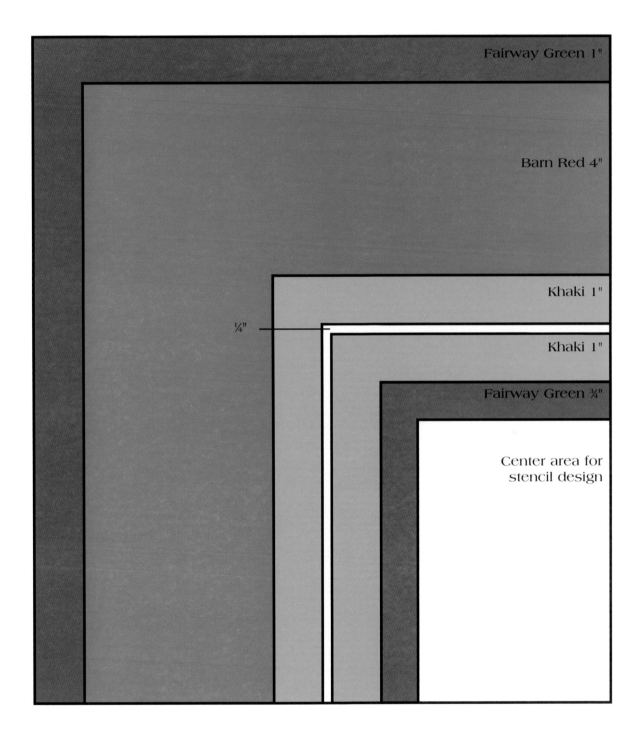

Fairway Green 1"

Barn Red 4"

Khaki 1"

¼"

Khaki 1"

Fairway Green ¾"

Center area for
stencil design

Fruit Medley
Instructions begin
on page 67

Fruit Medley

Pictured on page 66

Designed by
Avis Everett

This floorcloth is a great way to brighten up any corner of your kitchen.

Floorcloth size: 52" square

GATHER THESE SUPPLIES

Painting Surface:
Canvas or vinyl remnant,
   measuring 52" square

Paints & Coatings:
Indoor/outdoor gloss enamels:
   Dandelion Yellow
   Deep Purple
   Real Green
   Real Red
   Spring Green
Matte acrylic sealer

Tools:
Chalk pencil
Paint roller
Stencil brushes: ⅜"; ¼" (3)
Tape measure

Other Supplies:
Masking tape: 1"-wide
Single-overlay stencil:
   Fruit Medley

INSTRUCTIONS

Prepare:
1. Refer to Preparing a Floorcloth on page 15. Measure, cut, and prepare floorcloth.

2. Using paint roller, base-coat floorcloth with Dandelion Yellow paint. Let dry.

3. Using chalk pencil and tape measure, mark 1" from outside edge of floorcloth. This will be the distance between outer edge and stencil pattern.

Paint the Design:
1. Mask off fruit designs of Fruit Medley stencil. Allow space to paint in all green stems on fruit.

2. Using ⅜" and ¼" stencil brushes, stencil stems, leaves, and top and bottom borders with Real Green. Repeat on all four sides of floorcloth. Let dry. After all stems and leaves are painted, wash stencil carefully and remove tape from fruit designs. Let dry.

3. Mask off all leaves, stems, and top and bottom borders on stencil.

Fruit:
1. Using ¼" stencil brush, stencil apples and cherries with Real Red.

2. Using ¼" stencil brush, stencil pears with Spring Green.

3. Using ¼" stencil brush, stencil grapes with Deep Purple. Let dry.

Finish:
1. Refer to Varnishing a Floorcloth on page 27. Seal floorcloth with matte acrylic sealer.✽

Rose Swag

Pictured on page 68

Designed by
Jane Gauss

Victorian in style, this rose-themed floorcloth would be the perfect finishing touch for a powder room.

Floorcloth size: 2' x 3'

GATHER THESE SUPPLIES

Painting Surface:
Canvas or vinyl remnant,
   measuring 2' x 3'

Paints & Coatings:
Indoor/outdoor satin acrylic:
   Vanilla
Gel paints:
   Forest Shade
   Pink Blush
   Wood Rose
Matte acrylic sealer

Tools:
Chalk pencil
Sponge brush: 1"
Ruler
Stencil brushes: ½" (3)

Other Supplies:
Masking tape: 1"-wide
Single-overlay stencil:
   Rose Swag

INSTRUCTIONS

Prepare:
1. Refer to Preparing a Floorcloth on page 15. Measure, cut, and prepare floorcloth.
Continued on page 69

Rose Swag
Instructions begin
on page 67

Continued from page 67

2. Using 1" sponge brush, base-coat floorcloth with one to two coats of Vanilla. Let dry.

Paint the Design:
1. Using Rose Swag design and ½" stencil brush, stencil leaves with Forest Shade. Refer to photo on page 68 for placement.

2. With ½" stencil brushes, stencil roses with Pink Blush and Wood Rose.

3. Using chalk pencil and ruler, lightly draw a 6" circle in center of floorcloth for placement of center roses.

4. Choose one rose from swag to use for center group. Tape over cut out areas around selected rose.

5. Center rose over pencil mark and stencil with Pink Blush and Wood Rose. Let dry.

Finish:
1. Refer to Varnishing a Floor-cloth on page 27. Seal floor-cloth with matte acrylic sealer.✻

Cottage Rose

Pictured on page 70

Designed by
Avis Everett

Bring the look of a quaint, country cottage home with this lovely floorcloth.

Floorcloth size: 52" x 5'

GATHER THESE SUPPLIES

Painting Surface:
Canvas or vinyl remnant, measuring 52" x 5'

Paints & Coatings:
Indoor/outdoor gloss enamels:
   Dandelion Yellow
   Raspberry
   Real Green
   Real Red
   Spring Green
   White
Latex wall paint: White
Matte acrylic sealer

Tools:
Chalk pencil
Paint roller
Sponge brush: 1" (2)
Stencil brushes: ½" (5)
Tape measure
Yardstick

Other Supplies:
Masking tape: 2"-wide
Single-overlay stencils:
   Checkerboard
   Cottage Rose

INSTRUCTIONS

Prepare:
1. Refer to Preparing a Floor-cloth on page 15. Measure, cut, and prepare floorcloth.

2. Using paint roller, base-coat floorcloth with mixture of Spring Green enamel and White latex paint. Let dry.

Paint the Design:
1. Using Cottage Rose stencil, mask off leaf areas. Using ½" stencil brushes, stencil flowers alternating with Dandelion Yellow, Raspberry, and Real Red around floorcloth, using outside edge for placement.

2. After roses are painted, wash stencil. Remove masking tape from leaves. Using ½" stencil brush, stencil leaves with Real Green. Let dry.

3. Using chalk pencil and tape measure, mark 5" from outside edge of floorcloth. Mask along outside of line, creating outer edge of yellow border.

4. Measure and mark inside from line another 5". Mask along outside of line to create inner edge of yellow border, leaving a 5"-wide border.

5. Using paint roller, paint border with Dandelion Yellow paint. Let dry.

6. Remove masking tape from outer edge of yellow border. Leave inner edge tape in place. By masking background color in this 2"-wide section, the next interior stripe is created.

Continued on page 71

Cottage Rose
Instructions begin
on page 69

Continued from page 69

7. Using Checkerboard design and chalk pencil, place stencil against edge of existing masking tape border and draw a line on inside edge of stencil. Work toward the center. Mask along pencil lines toward center.

8. Using paint roller, paint inside border with White indoor/outdoor paint. Let dry.

9. Using ½" stencil brush, stencil Checkerboard design on white border with Real Green. Let dry.

10. Remove masking tape from outer edge of checkerboard border. Leave inside edge tape in place. By masking background color in this 2"-wide section, you will be creating the next interior stripe.

11. Using 1" sponge brush, paint center of floorcloth with two coats of Raspberry. Let dry.

12. Remove all tape. Using yardstick and 1" sponge brush, make corrections on borderlines. Let dry.

Finish:

1. Refer to Varnishing a Floorcloth on page 27. Seal floorcloth with matte acrylic sealer.✳

Noah's Ark

Pictured on page 72

Designed by
Jane Gauss

With a theme that never goes out of style, this floorcloth adds a distinctive, custom look to any child's room.

Floorcloth size: 2' x 3'

GATHER THESE SUPPLIES

Painting Surface:
Canvas or vinyl remnant, measuring 2' x 3'

Paints & Coatings:
Dry brush stencil paints:
  Andiron Black
  Berry Red
  Forest Shade
  Gold Metallic
  Shadow Gray
  Ship's Fleet Navy
  Sunny Brooke Yellow
  Truffles Brown
  Wildflower Honey
Faux-finishing glazes:
  Danish Blue
  Neutral
  Sage Green
Latex wall paint: White
Matte acrylic sealer

Tools:
Chalk pencil
Scissors
Sponging mitt
Stencil brushes: ½" (9)
Tape measure

Other Supplies:
Cardboard for template
Disposable plates
Masking tape: 1"-wide

Single-overlay stencils:
  Basket Weave, 2"-wide
  Noah's Ark

INSTRUCTIONS

Prepare:
1. Refer to Preparing a Floorcloth on page 15. Measure, cut, and prepare floorcloth. There is no need to base-coat floorcloth since prime coat is white.

2. Using chalk pencil and tape measure, mark 4"-wide border on outer edges of floorcloth. Mask off along mark.

3. Using disposable plate, mix Neutral glaze with Danish Blue glaze to achieve a dark tint. Dampen a sponging mitt, then sponge glaze mixture on outer border. Remove tape. Let dry.

4. Mask off a 2"-wide border on inside of outer border. Using Basket Weave stencil and ½" stencil brush, stencil design inside border with Gold Metallic. Remove tape. Let dry.

5. Mask off remaining central section of floorcloth. Using scissors, create cardboard template to act as horizontal line between sky and hills. Refer to photo on page 70 for placement.

Paint the Design:
1. Using disposable plate, mix Neutral glaze with Danish Blue glaze to achieve a light tint. With cardboard template in place, lightly sponge glaze over sky area. Let dry.
Continued on page 73

Noah's Ark
Instructions begin
on page 71

Continued from page 71

2. Using disposable plate, mix Neutral glaze with Sage Green glaze to achieve a light tint. Sponge mixture over grass and hills area, be certain there are no white gaps between sky and hills.

3. Using disposable plate, mix Neutral glaze with Sage Green glaze to create darker tint than mixture used in Step 2. Sponge mixture on grass area, using curved sections of cardboard template, to create darker hills among grass. Let dry.

4. Using ½" stencil brushes, stencil ark with Berry Red, Shadow Gray, and Ship's Fleet Navy. Refer to photo on page 72 for placement.

5. Stencil ducks with Berry Red and Sunny Brooke Yellow.

6. Stencil alligators with Forest Shade.

7. Stencil elephants with Shadow Gray.

8. Stencil giraffes with Truffles Brown and Wildflower Honey.

9. Stencil sheep with Andiron Black and Shadow Gray.

10. Add touches of Sunny Brooke Yellow to grass areas and to pounce in sun. Let dry.

Finish:
1. Refer to Varnishing a Floor-cloth on page 27. Seal floor-cloth with matte acrylic sealer.✳

Fresh Herbs

Pictured on page 74

Designed by
Stewart Huntington

Delicate, free-flowing herbs bring this floorcloth to life. After completing floorcloth, stencil matching placemats and napkins.

Floorcloth size: your choice

GATHER THESE SUPPLIES

Painting Surface:
Canvas or vinyl remnant, measuring your choice

Paints & Coatings:
Acrylic craft paints:
  Butter Pecan
  Poetry Green
Dry brush stencil paints:
  English Lavender
  Herb Garden Green
  Sage Green
  Sherwood Forest Green
  Sunny Brooke Yellow
  Wild Ivy Green
Matte acrylic sealer

Tools:
Chalk pencil
French brush
Stencil brushes: ½" (6)
Tape measure

Other Supplies:
Masking tape: ½"-wide
Paper towels
Stencil: Cut your own stencil

INSTRUCTIONS

Prepare:
1. Refer to Preparing a Floor-cloth on page 15. Measure, cut, and prepare floorcloth.

2. Using chalk pencil and tape measure, mask off center area of floorcloth.

3. Dip French brush in Poetry Green. Blot brush on paper towel until almost dry. Brush on floorcloth in random diagonal strokes. Let some white background show through.

4. Repeat step 2, using Butter Pecan. Let dry.

5. Mask off outer and inner borders. Paint with Poetry Green. Let dry 30 minutes. Remove tape. Let dry.

6. Mask off border outside inner Poetry Green border. Paint with Butter Pecan. Let dry 30 minutes. Remove tape. Let dry.

7. Refer to Cutting a Stencil on page 75. Using Fresh Herbs Patterns on page 76, cut your own stencil.

Paint the Design:
1. Using thyme stencil, create border around outer edge of floorcloth.

2. Using ½" stencil brushes, stencil rosemary, sage, and thyme flowers with English Lavender. Stencil parsley flowers with Sunny Brooke Yellow.

3. Stencil thyme leaves with Wild Ivy Green; sage leaves with Sage Green; rosemary leaves with Sherwood Forest Green; and parsley leaves with Herb Garden Green.

Finish:
1. Refer to Varnishing a Floor-cloth on page 27. Seal floor-cloth with matte acrylic sealer.✳

Fresh Herbs
Instructions begin
on page 73

Cutting a Stencil

STENCIL-CUTTING SUPPLIES

• Craft knife (pencil-style) with #11 blades—for cutting stencil blank sheets.

• Glass cutting board, ⅜"- or ½"-thick with ground or sanded edges—for cutting surface.

• Permanent, fine-tip marker—for writing on stencil material.

• Stencil blank sheets—for creating stencils.

• White paper—for placement under glass. *Options*: Buy frosted glass or put glass on a light-box.

CREATING YOUR OWN PATTERN

When creating your own pattern, keep in mind the colors you want to use and the fact that stencil designs must have bridges to keep them strong once you cut them. Outline all edges of the design to be cut, using a fine-tip marker. (You may find it less confusing to color the drawing with colored pencils at this point to help see the design and separate the colors for making multiple overlays.) Place stencil blank sheet over your design. Trace the first color you will be cutting with solid lines on the stencil sheet. Trace lines for other colored overlays as dashed or dotted lines. Use a separate sheet of stencil blank sheet for each overlay (one overlay per color).

USING EXISTING PATTERNS

Using a photocopier, reduce or enlarge patterns as needed. Place stencil blank sheet over design. Trace the first color you will be cutting with solid lines on stencil material. Trace lines for other colored overlays as dashed or dotted lines. Use a separate sheet of stencil blank material for each overlay (one overlay per color).

1 Create pattern and outline. Color pattern.

2 Trace pattern.

CUTTING STENCILS

Place white paper under glass cutting surface. Put traced design on stencil blank material then on top of glass. Cut along lines, holding craft knife like a pencil.

Cutting Tips:
• Hold craft knife with blade facing you. Pull knife toward you to cut. Keep fingers out of path of knife.

• Use noncutting hand to hold and move stencil blank material. Hold cutting blade stationary.

• Let shaft of knife roll in fingers to help go around corners.

• Cut out small, intricate shapes first.

• Try to cut out shape without lifting blade from stencil blank material. Exceptions to this are pointed shapes with lots of edges.

• Use sufficient pressure to cut through stencil material with one stroke—this will prevent a jagged edge that can result from having to go back over cut. Remember the edge you cut is the edge you will paint.

Fresh Herbs Patterns

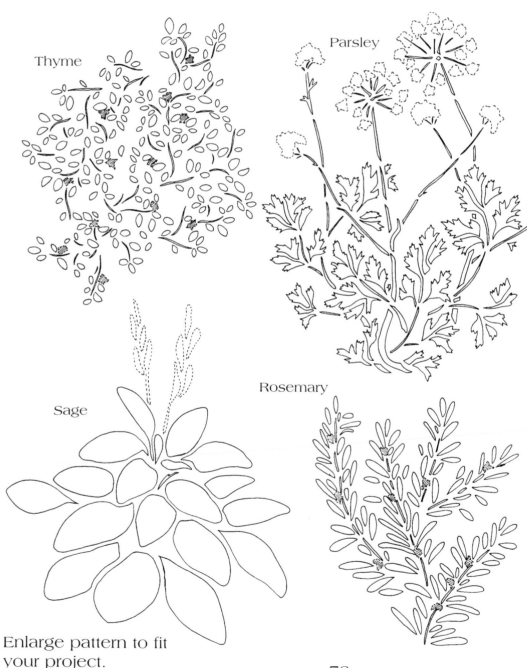

Thyme

Parsley

Sage

Rosemary

Enlarge pattern to fit your project.

Découpaged
Floorcloths

A favorite print, a favorite fabric, or any type of paper design can be découpaged onto a floorcloth to grace your floors with designs you love. In this chapter, you will learn how easy découpage really is! There is a poinsettia floorcloth for the holidays and a game fish design, which combines découpage with stenciling.

Découpaging

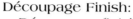

DÉCOUPAGING SUPPLIES

Acrylic Craft Paints:

Acrylic craft paints are high-quality paints perfect for base-coating floorcloth surfaces. They are available in a wide array of decorator colors and cleanup is easy with soap and water.

Découpage Finish:

Découpage finish is a nontoxic, water-based glue and sealer in one. It dries quickly with a hard finish that can be sanded or polished. This product cleans up easily with soap and water, and is available in a clear finish or an antique finish, which imparts a slightly yellowed color for instant aging of projects.

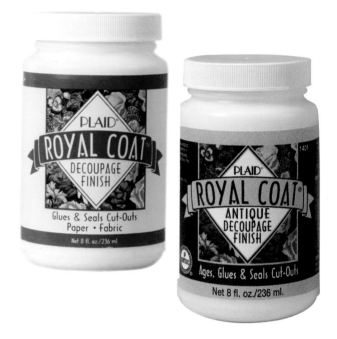

Brushes:

Sponge brushes are suggested for applying découpage finish. The size of brush used will depend upon the size of the working surface.

Cutting Tools:

A pair of small, sharp-pointed découpage scissors is essential for success in cutting out paper designs. A pencil-style craft knife with a sharp #11 blade and a cutting board are useful for cutting out inside pieces of motifs and for large pieces of découpage paper. For cutting out fabric designs, use a sharp pair of embroidery scissors, which are scissors specifically designed for cutting in detailed areas.

Fabric Cutouts:

Fabric, as well as paper, can be used for découpaging. The variety of motifs and patterns available at your local fabric store is enormous. Be certain to wash fabric first, following manufacturer's instructions, then press well with iron.

Paper Cutouts:

Paper cutouts are most often used in découpaging. Wallpaper, wrapping paper, and antique prints are just some of the types of patterned paper used. You can also use color copies or hand-colored black-and-white copies of prints. Be certain to choose a weight of paper that will not buckle when découpage finish is applied.

Découpaging

1 Using découpage or embroidery scissors, cut out your design. If you choose to add a decorative edge to your paper print, you can tear the image or burn the torn edges with a candle.

2 Using a sponge brush, apply découpage finish to the back of the print or fabric and adhere it onto the project surface, smoothing the print with your fingers.

Découpaged poinsettia

3 Apply two coats of découpage finish to project. Let dry 20 minutes between coats. For a high-gloss finish, allow the project to cure for one week, then spray project with a gloss acrylic sealer.

Christmas
Poinsettia
Instructions begin
on page 81

Christmas Poinsettia

Pictured on page 80

Greet guests at Christmastime with this elegant poinsettia floorcloth. Use these same techniques for creating a Christmas tree skirt!

Floorcloth size: 2' x 3'

GATHER THESE SUPPLIES

Painting Surface:
Canvas or vinyl remnant,
    measuring 2' x 3'

Paints & Coatings:
Acrylic craft paint:
    Pure Gold Metallic
Découpage finish
Matte acrylic sealer

Tools:
Chalk pencil
Sponge brushes: ½" (2)
Iron
Paint roller
Scissors: embroidery; fabric
Tape measure

Other Supplies:
Fabric: black fabric with gold
    and poinsettia accents
Masking tape: ½"-wide

INSTRUCTIONS

Prepare:
1. Refer to Preparing a Floorcloth on page 15. Measure, cut, and prepare floorcloth.

2. Using chalk pencil and tape measure, mark 4" in from outside edge of floorcloth.

3. From this line, measure and mark a ½"-wide border towards inside of floorcloth. Mask these two chalked lines with tape so ½" border is exposed.

Paint the Design:
1. Using paint roller, paint ½"-wide border with Pure Gold. Let dry.

2. Remove masking tape. Using fabric scissors, cut fabric for outside border in 5"-wide strips. Refer to photo on page 80 for placement of mitered edges. Cut strips of 5"-wide fabric into lengths slightly longer than sides, leaving ¼" on either end. Using iron, press and turn under fabric at mitered edges.

3. Press inside edge of strip under ½". For outer edge, turn under additional ½".

4. Using ½" sponge brush, adhere fabric with découpage finish. Be certain to smooth fabric with fingers to avoid bumps. Join mitered edges.

5. Using embroidery scissors, cut several large poinsettias with leaves from additional fabric.

6. Découpage large poinsettias to center area of floorcloth, referring to photo for placement. Notice poinsettias in photo are overlapped in some areas.

Finish:
1. Refer to Varnishing a Floorcloth on page 27. Apply two coats of découpage finish to fabric and poinsettias. Let dry 20 minutes between coats.❈

Game Fish
Instructions begin
on page 83

Game Fish

Pictured on page 82

Designed by
Jane Gauss

This is the one that did not get away. This combination of découpage and stenciling makes a fun floorcloth.

Floorcloth size: 24" half circle

GATHER THESE SUPPLIES

Painting Surface:
Canvas or vinyl remnant,
 measuring 24" half circle

Paints & Coatings:
Acrylic craft paint:
 Tapioca
Découpage finish
Dry brush stencil paints:
 Romantic Rose
 Sherwood Forest Green
 Wild Ivy Green
Faux-finishing glazes:
 Baby Pink
 Italian Sage
 Moss Green
 Neutral
Matte acrylic sealer

Tools:
Fabric scissors
Paint roller
Sponge brush: ½"
Sponging mitt or sea sponge
Stencil brush: ½"

Other Supplies:
Disposable plates
Fabric: picture panel of a
 game fish
Single-overlay stencil:
 Classic Ivy

INSTRUCTIONS

Prepare:
1. Refer to Preparing a Floorcloth on page 15. Measure, cut, and prepare floorcloth.

2. Using paint roller, base-coat floorcloth with Tapioca. Let dry.

Paint the Design:
1. Using disposable plate, mix Italian Sage with Neutral to obtain a light tint.

2. Using sponging mitt, sponge mixture over entire floorcloth. Repeat with mixture of Moss Green and Neutral. Repeat with a mixture of Baby Pink and Neutral.

3. Using fabric scissors, cut fabric fish from picture panel and place in center of floorcloth. Refer to photo on page 82 for placement.

4. Using ½" sponge brush, adhere fabric cutout with découpage finish. Let fabric applique dry overnight.

5. Seal top of fabric fish with découpage finish. Let dry.

6. Using Classic Ivy stencil and ½" stencil brush, stencil ivy around outside edge of floorcloth with Wild Ivy Green. Randomly position ivy to go around edges.

7. Shade with Sherwood Forest Green. Highlight toward tips of some leaves with Romantic Rose. Let dry.

Finish:
1. Refer to Varnishing a Floorcloth on page 27. Seal floorcloth with matte acrylic sealer.✳

Decorative Painted Floorcloths

This chapter showcases three fun floorcloths, using the technique of decorative painting. With the patterns provided, you can create custom-looking floorcloths with nothing more than paints, brushes, and your own artistic talent.

Decorative Painting

DECORATIVE PAINTING SUPPLIES

Acrylic Craft Paints:

Acrylic craft paints are premixed, quick-drying paints that are available in an extensive array of colors. They are excellent for decorative painting as well as stenciling. It is also easy to mix custom colors to coordinate with home decor. Cleanup is easy with soap and water.

Artist's Brushes:

Artist's brushes are high-quality brushes with either nylon or natural bristles, wooden handles, and metal ferrules. They come in a variety of sizes and styles, and are perfect for decorative painting. The size and type of brush depend on the size of project to be painted and the type of strokes implemented.

Other Supplies:

• Tracing paper and pencil—for tracing patterns.

• Transfer paper and stylus—for transferring pattern onto floorcloth.

Sunflower
Garden
Instructions begin
on page 87

Sunflower Garden

Pictured on page 86

Designed by
Tasha Yates

Bright colors and a whimsical nature design combine to make this floorcloth fun.

Floorcloth size: 2' x 3'

GATHER THESE SUPPLIES

Painting Surface:
Canvas or vinyl remnant, measuring 2' x 3'

Paints & Coatings:
Acrylic craft paints:
Asphaltum
Burgundy
Hauser Green Dark
Licorice
Raspberry Wine
Tapioca
Tartan Green
Teddy Bear Tan
Turner's Yellow
Blending medium
Matte acrylic sealer

Tools:
Chalk pencil
Flat brush: #10
Liner brush
Paint roller
Round brush: #1
Shaders: #2; #4; #8
Stencil brush: 1"

Other Supplies:
Disposable plate
Extrafine-tip permanent marker, black

Masking tape
Stencil: Cut your own stencil
Tracing paper and pencil
Transfer paper and stylus

INSTRUCTIONS

Prepare:
1. Refer to Preparing a Floorcloth on page 15. Measure, cut, and prepare floorcloth.

2. Using paint roller, base-coat floorcloth with Tapioca.

3. Using disposable plate, mix equal parts Tapioca and Teddy Bear Tan with a small amount of blending medium.

4. Using 1" stencil brush, apply paint to floorcloth with a light, swirling motion in alternating directions. Start at edge of floorcloth and work inward, using more pressure along edges and less pressure in center area, creating a mottled effect. Add more Teddy Bear Tan to make edges darker. Let dry.

5. Refer to Cutting a Stencil on page 75. Using tracing and transfer tools, transfer Sunflower Garden Patterns on pages 88–89 onto stencil material.

Paint the Design:
1. Using chalk pencil, draw a line above bottom checkerboard border for grass area.

2. Using #10 flat brush, paint grass with equal parts Tartan Green and blending medium.

3. Paint alternating checks in border with Licorice.

4. Paint sunflower centers with Licorice. Let dry.

5. Paint sunflower petals with Turner's Yellow. Using #4 shader, shade with Asphaltum.

6. Using liner brush, paint sunflower stems with Asphaltum. Shade with equal parts Asphaltum and Licorice.

7. Using #10 flat brush, paint sunflower leaves with Tartan Green. Shade with Hauser Green Dark. Deepen shadows with equal parts Hauser Green Dark and Licorice.

8. Using handle end of #10 flat brush, add Tapioca dots to centers.

9. Base-coat watering can with Burgundy. Shade with Raspberry Wine. Deepen shadows with equal parts Licorice and Raspberry Wine.

10. Paint rim of watering can spout with Tartan Green. Shade with Hauser Green Dark.

11. Paint end of spout with Turner's Yellow. Shade with Asphaltum.

12. Base-coat angel's dress with Burgundy. Shade with Raspberry Wine. Deepen shadows with equal parts Licorice and Raspberry Wine.

13. Paint angel's hair and star with Turner's Yellow. Shade with Asphaltum.

14. Paint hands, feet, face, and wings of angel with Tapioca. Shade with Teddy Bear Tan.

15. Base-coat bunny with Teddy Bear Tan. Shade with Asphaltum.

16. Using #2 shader, highlight bunny with Tapioca.

17. Using liner brush, paint face and whiskers with Asphaltum.

18. Paint some of the small flower circles with Burgundy and others with Turner's Yellow.

19. Using # 2 shader, shade Turner's Yellow flowers with Asphaltum. Shade Burgundy flowers with Raspberry Wine.

20. Paint leaves of small circle flowers with Tartan Green. Shade with Hauser Green Dark.

21. Using liner brush, paint stems with Asphaltum.

22. Paint sprigs of grass with Hauser Green Dark.

23. Using extrafine-tip permanent marker, outline designs. Let dry.

Finish:
1. Refer to Varnishing a Floor-cloth on page 27. Seal floor-cloth with matte acrylic sealer.✳

Sunflower Garden Patterns

Enlarge pattern 250%

Sunflower Garden Patterns

Enlarge patterns 250%

Ribbon & Holly
Instructions begin
on page 91

Ribbon & Holly

Pictured on page 90

Designed by
Donna Dewberry

This floorcloth showcases a beautiful bough of holly and a border to paint your own favorite holiday greeting.

Floorcloth size: 36" x 42"

GATHER THESE SUPPLIES

Painting Surface:
Canvas or vinyl remnant, measuring 36" x 42"

Paints & Coatings:
Acrylic craft paints:
  Berry Wine
  Green Forest
  Holiday Red
  Inca Gold Metallic
  Maple Syrup
  Sunflower
  Wicker White
Matte acrylic sealer

Tools:
Flat brushes: ¾"; #10
Paint roller
Script liner: #2
Sponging mitt or sea sponge

Other Supplies:
Chalk pencil
Masking tape: 1"-wide
Old toothbrush
Tape measure
Tracing paper and pencil
Transfer paper and stylus

INSTRUCTIONS

Prepare:

1. Refer to Preparing a Floorcloth on page 15. Measure, cut, and prepare floorcloth.

2. Using paint roller, base-coat floorcloth with two coats of Wicker White. Let dry between coats.

3. Using chalk pencil and tape measure, mark 6"-wide border from outside edge of floorcloth. Mask off remaining center area.

Paint the Design:

1. Using damp sponging mitt, pounce center area of floorcloth with Maple Syrup. Apply paint so edges are a little darker than center. Let dry. Remove tape.

2. Mask off a 1"-wide border around outer edge of floorcloth and a 5"-wide border within outer border.

3. Using sponging mitt, apply Berry Wine in a circular motion around 5" border. Let dry. Remove tape.

4. Mask off outer, 1" border. Using ¾" flat brush, apply Green Forest to 1" outer edge. Remove tape. Let dry.

5. Paint 1" squares of Maple Syrup at each corner.

6. Paint 1" squares of Inca Gold Metallic on each side of each Maple Syrup square.

7. Using #2 script liner, add accent X's on each Maple Syrup corner with Inca Gold Metallic.

8. Using chisel edge of ¾" flat brush, paint an accent line around inside border with Green Forest. Let dry.

9. Using tracing and transfer tools, transfer Ribbon & Holly Pattern on page 94 onto center of floorcloth.

10. Refer to Ribbon & Holly Worksheet on page 93. Double-load ¾" flat brush with Berry Wine and Holiday Red to paint ribbons. Refer to the One Stroke™ Painting Technique on page 92 for information on double-loading.

11. Double-load ¾" flat brush with Green Forest and Sunflower to paint holly.

12. Using chisel edge, paint pine needles with Green Forest and Sunflower. Make short strokes, leading with Sunflower.

13. Using handle end of brush, dot berries on holly with Berry Wine.

14. Using #2 script liner, add accents to ribbon with Inca Gold Metallic.

15. Using ¾" brush, paint letters around outer border with Inca Gold Metallic. Let dry.

16. Using old toothbrush, spatter project surface with thinned Maple Syrup. Let dry.

Finish:

1. Refer to Varnishing a Floorcloth on page 27. Seal floorcloth with matte acrylic sealer.❋

One Stroke™ Painting Technique

1 For making most strokes, you will use the chisel edge of a flat brush.

2 To load a brush, dip one corner of the brush into the first color. Dip the other corner of the brush into second color. The colors should form triangles on the corners of the brush and meet in the middle.

3 With the multiloaded brush, you can paint many designs with just one or two strokes.

Ribbon & Holly Worksheet

Ribbon
Begin on chisel edge.

Stripes
Use script liner with Wicker White or Inca Gold.

Chisel edge.

Smaller Leaves
Load #12 flat brush with Green Forest and Sunflower.

Holly
Double-load ¾" flat brush with Green Forest and Sunflower.

Paint first half.

Paint second half.

Using chisel edge, add stem.

Berries
Using handle end of brush, dot berries with Holiday Red and Inca Gold.

Enlarge pattern 200%

Jingle Bells
Santa
Instructions
begin on
page 96

Jingle Bells Santa

Pictured on page 95

Designed by
Donna Dewberry

Brighten your home at holiday time with this festive floorcloth.

Floorcloth size: 38" x 52½"

GATHER THESE SUPPLIES

Painting Surface:
Canvas or vinyl remnant, measuring 38" x 52½"

Paints & Coatings:
Acrylic craft paints:
    Berry Wine
    Burnt Umber
    Grass Green
    Green Forest
    Inca Gold Metallic
    Midnight
    Napthol Crimson
    Rose Chiffon
    Skintone
    Sunflower
    Wicker White
Matte acrylic sealer

Tools:
Chalk pencil
Flat brushes: ¾"; #6; #12
Mini scruffy brush
Script liner: #2
Scruffy brush
Tape measure

Other Supplies:
Masking tape: 1"-wide
Old toothbrush
Tracing paper & pencil
Transfer paper & stylus

INSTRUCTIONS

Prepare:

1. Refer to Preparing a Floorcloth on page 15. Measure, cut, and prepare floorcloth.

2. Using chalk pencil and tape measure, mark 8" squares at all four corners of floorcloth.

3. Using #12 flat brush, basecoat squares with Inca Gold Metallic. Remove tape carefully. Let dry.

4. Mask off spaces between gold squares. Be certain edges of tape are securely adhered.

5. Base-coat with Napthol Crimson. Remove tape. Let dry.

6. Mask off center area. Be certain edges of tape are securely adhered.

7. Base-coat with Green Forest. Remove tape. Let dry.

8. Using tracing and transfer tools, transfer Jingle Bells Santa Patterns on pages 97–98 onto floorcloth.

Paint the Design:

1. Refer to One Stroke™ Painting Technique on page 92. Double-load ¾" flat brush with Rose Chiffon and Skintone.

2. Refer to Jingle Bells Santa Face Worksheet on page 99. Base-coat face, using even strokes.

3. Double-load ¾" flat brush with Berry Wine and Napthol Crimson. Paint cap in overlapping strokes with Berry Wine toward outer edge.

4. Load ¾" flat brush with Wicker White. Sideload with Burnt Umber. Paint beard, stroking downward on chisel edge of brush, leading with Wicker White.

5. Double-load #12 flat brush with Napthol Crimson and Rose Chiffon. Paint nose with reverse U-stroke, with Napthol Crimson toward top. Paint lip with a U-stroke. Let dry.

6. Load chisel edge of #6 flat brush with Wicker White. Paint mustache. Sideload same brush with Burnt Umber. Paint detailing with chisel edge of brush.

7. Load #12 flat brush with Wicker White. Paint eyeballs. Let dry.

8. Load #6 flat brush with Midnight. Paint pupils. Let dry.

9. Double-load mini scruffy brush with Burnt Umber and Wicker White. Pounce eyebrows, using choppy, up and down strokes.

10. Double-load scruffy brush with Burnt Umber and Wicker White. Pounce fur band and pompom on cap, keeping Burnt Umber side of brush on upper part of band and outer part of pompom.

11. Load #2 script liner with inky Burnt Umber. Using tip of brush, outline eyes and paint eyelashes.

12. Load #2 script liner with inky Wicker White. Using tip of brush, paint highlights and twinkles in eyes. Let dry.

13. Using tip of brush, paint stars and outline cap with Inca Gold Metallic.

14. Refer to Ribbon & Holly Worksheet on page 93. Double-load ¾" flat brush with Green Forest and Sunflower. Paint holly leaves.

15. Double-load ¾" flat brush with Green Forest and Sunflower. Paint pine needles with chisel edge of brush.

16. Using handle end of ¾" flat brush, dot berries with Napthol Crimson. Let dry.

17. Load #2 script liner with Inca Gold Metallic. Using tip of brush, highlight pine needles and berries.

18. Double-load #6 flat brush with Green Forest and Grass Green. Using flat edge of brush, paint lettering. Let dry.

19. Load #6 flat brush with Inca Gold Metallic. Highlight lettering. Let dry.

20. Using old toothbrush, spatter floorcloth with inky Inca Gold Metallic. Let dry.

Finish:
1. Refer to Varnishing a Floorcloth on page 27. Seal floorcloth with matte acrylic sealer.✳

Jingle Bells Santa Pattern

Enlarge pattern 250%

Enlarge pattern 250%

Jingle Bells Santa Face Worksheet

Face
Double-load ¾" flat brush with Rose
Chiffon and Skintone.

Mustache
Load chisel edge of #6 flat brush
with Burnt Umber and Wicker White.

Eyeballs
Paint eyeballs with
Wicker White.

Pupils
Using #6 flat brush, paint
pupils with Midnight.
Stroke first half, then
second.

Eyelids
Using #1 script liner, paint lines
around eyes with inky Burnt
Umber.

Eyelashes
Using #2 script
liner, paint
eyelashes with
inky Burnt Umber.

Beard
Paint beard
before mustache.
Double-load
chisel edge of ¾"
flat brush with
Burnt Umber and
Wicker White.
Stroke beard.

Eyebrows
Double-load mini scruffy brush
with Burnt Umber and Wicker
White. Pounce eyebrows.

Star
Load #2 script
liner with Inca
Gold Metallic.

99

Stamped Floorcloths

It is as simple as loading a stamp with paint and stamping a design on a straightedged floor-cloth. It is addictively fun. In this chapter, you will learn to use stamps and blocks to create a variety of designs. Stamps and blocks are so easy to use; the only limit is your imagination.

Stamping & Block Printing

STAMPING & BLOCK PRINTING SUPPLIES

Printing Blocks:

Printing blocks are soft pieces of die-cut, pliable material used to create solid designs with paint. They are available in a wide variety of designs and because they are so durable, they can be used again and again. Cleanup is easy with gel cleaner and water.

Stamping Blocks:

Stamping blocks are easy to use, flexible stamps made from a dense, pliable material suitable for stamping on a variety of surfaces.

Stamps:

Stamps differ from blocks in that they usually include some interior detail, such as the veins of a leaf; whereas, blocks are solid shapes with no interior detail. Stamps are available in a wide range of shapes, are durable for long-term use, and cleanup easily with gel cleaner and water.

Other Supplies:
• Tracing paper & pencil—for tracing patterns.

• Transfer paper & stylus—for transferring pattern onto floorcloth.

Tools For Loading Blocks and Stamps:
• Artist's brushes—for loading paint on stamps and blocks. Loading with a brush is especially useful when loading different colors on different parts of a stamp or block.

• Glaze rollers—for providing the fastest and most consistent way to load a stamp or block.

• Sponge pouncing tool—for applying paint to stamps and blocks with ease and control.

• Sponge wedges—for applying paint to stamps and blocks. These dense pieces of sponge are easy to control and to clean up.

101

STAMPING AND BLOCK PRINTING PAINTS

Acrylic Craft Paints:

Acrylic craft paints are premixed, quick-drying paints that are available in an extensive array of colors. They are excellent for decorative painting as well as stenciling, stamping, and block printing. It is also easy to mix your own custom colors to coordinate with home decor. Cleanup is easy with soap and water.

Indoor/outdoor Satin Acrylics:

Indoor/outdoor satin acrylics are available in a wide variety of beautiful colors with a satin finish. These paints are not only durable on surfaces that receive a lot of wear, such as floors, but they stand up to the weather as well. These are your best choice for projects that will be used outdoors. No sealing is needed for these sturdy paints.

Faux-finishing Glazes:

Faux-finishing glazes are rich, thick mediums with a subtle transparency that are perfect for stamping, block printing, and faux finishing. They are available in a wide range of decorator colors, water-based, and nontoxic. Cleanup is simple with soap and water. Neutral glaze can be mixed with glaze colors to create an assortment of tints for textured effects.

Other Supplies:

• Gel cleaner—for cleaning stamping and block-printing tools. Make certain to clean tools before paint dries.

• Mitts and combs—for adding texture to all kind of surfaces.

Stamping

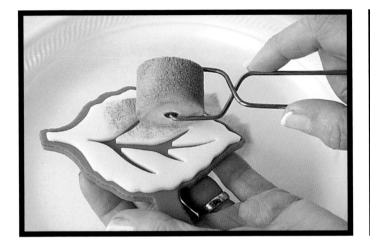

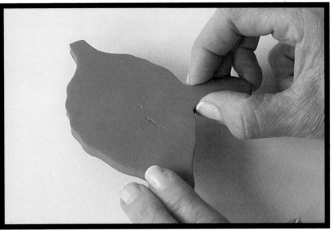

1 Squeeze a dime-sized amount of glaze on a disposable plate. Hold stamp by handle. With roller, apply a thin coat of glaze to cut side of stamp. Roll glaze to edges of block. Avoid getting glaze on the handle.

2 Release handle and gently press on back of stamp, pressing first in the center of the design, then on the edges. Use your fingers, not the heel of your hand, for consistent pressure in all areas.

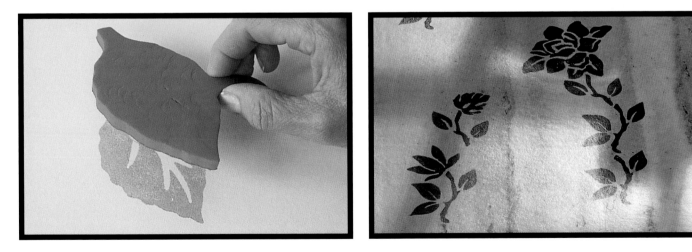

3 Lift stamp off surface, using the handle.

Magnolia stamp

Block Printing

1 Squeeze a dime-sized amount of glaze on a disposable plate. Hold block by handle. With flat brush, apply a thin coat of glaze to cut side of block. Brush glaze out to edges of block. Avoid getting glaze on the handle.

2 Holding loaded block by the handle, place on project surface. Release handle. Gently press block against surface without sliding it. On smaller blocks, walk fingertips over surface of block. On larger blocks, press with palm of hand. Be certain to press all edges.

3 Lift block off surface, using the handle. Move to another area and make a second press, repeating until it is necessary to reload. You can get two to five presses, depending on how intense you want the color.

Fruits and Casablanca Blocks

Marble &
Acanthus
Instructions begin
on page 107

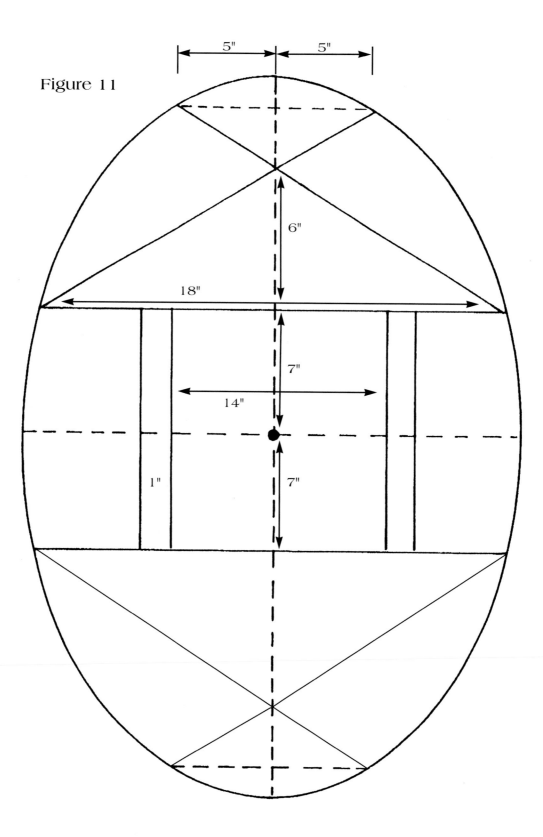

Figure 11

Marble & Acanthus

Pictured on page 105

Designed by
Kathi Malarchuk

A combination of stamping and marbleizing give this floorcloth a classic, elegant appeal.

Floorcloth size: 28" x 36" oval

GATHER THESE SUPPLIES

Painting Surface:
Canvas or vinyl remnant, measuring 28" x 36" oval

Paints & Coatings:
Faux-finishing glaze: Neutral
Indoor/outdoor satin acrylics:
  Caramel
  Chamois
  Pediment
  Twig
  White

Tools:
Chalk pencil
Flat brush: 2"
French brush
Marbleizing feather
Sponge brush: 1"
Tape measure

Other Supplies:
Grout tape: ¼"
Rag
Stamp: Acanthus Leaf

INSTRUCTIONS

Prepare:
1. Refer to Preparing a Floorcloth on page 15. Measure, cut, and prepare floorcloth.

2. Using 1" sponge brush, base-coat floorcloth with Chamois. Let cure for 72 hours.

3. Using chalk pencil and tape measure, mark vertical and horizontal centerlines of floorcloth. Refer to Figure 11 on page 106.

4. Using ¼" grout tape, mask off grout lines. The 28" vertical lines and 14" horizontal lines will create a 14" square in floorcloth center.

5. Place a mark 5" up and down from center horizontal line on outside edges.

6. Place a mark 6" to right and left of square on horizontal centerline.

7. With grout tape, mask off the X on each side of floorcloth, using the 6" mark as the center, the 5" mark as the outer ends of each X, and the top and bottom of the 28" vertical lines as the inner ends of each X.

Marbleizing:
Note: When marbleizing, complete one segment at a time so paint does not dry while working on another segment. Vary direction of "drifts" in each different segment to add interest to overall pattern.

1. Using 1" sponge brush, apply a light coat of Neutral glaze over segment.

2. While still wet, use French brush to lightly pounce drifts of White at a 30° angle, allowing background color to show through.

3. Wipe off brush on rag. Repeat with Pediment, then with Caramel.

4. Using 2" flat brush, blend and soften drifts by pouncing over paint surface.

5. While paint is still wet, dip marbleizing feather tip in Twig and create veins by pulling and pushing feather edge; following edges of drifts.

6. Repeat marbleizing in each segment, varying directions of drifts and amounts of colors.

7. Let dry. Remove grout tape.

Finish:
1. Refer to photo on page 105 for placement. Using Acanthus Leaf design, stamp around outer edge of floorcloth with Twig.

2. Let dry for 72 hours. There is no need to seal floorcloth when using indoor/outdoor satin acrylics. ✳

Magnolia Silhouettes
Instructions begin
on page 109

Magnolia Silhouettes

Pictured on page 108

Designed by
Susan Goans Driggers

The soft and natural background of this floorcloth make the magnolia stamps stand out.

Floorcloth size: 2' x 3'

GATHER THESE SUPPLIES

Painting Surface:
Canvas or vinyl remnant,
    measuring 2' x 3'

Paints & Coatings:
Faux-finishing glaze: Neutral
Indoor/outdoor satin acrylics:
    Black
    Caramel
    Espresso
    Limestone

Tools:
French brush
Old toothbrush
Paint roller

Other Supplies:
Containers (3)
Damp cloth
Spray bottle filled with water
Stamps: Magnolia Collection

INSTRUCTIONS

Prepare:
1. Refer to Preparing a Floorcloth on page 15. Measure, cut, and prepare floorcloth.

2. Using paint roller, base-coat floorcloth with Limestone. Let dry.

3. Mix Caramel, Espresso, and Limestone in separate containers with Neutral glaze.

Paint the Design:
1. Dip tip of French brush into Limestone mixture and stroke lightly onto entire surface.

2. Dip French brush into Caramel mixture. While Limestone paint is still wet, pounce surface with streaks of Caramel. Work pattern in a linear way.

3. Clean bristles by wiping on a damp cloth. Using spray bottle, mist surface to keep moist.

4. Dip French brush into Espresso mixture and pounce and brush light streaks onto surface.

5. Using old toothbrush, spatter surface lightly with Espresso that has been diluted with water. Let dry.

6. Using Magnolia Collection, design stamp magnolia onto dry floorcloth with Black.

Finish:
1. Let floorcloth dry 72 hours. With indoor/outdoor satin acrylics, there is no need to seal floorcloth.✳

Forest Floor

Pictured on page 110

Design © by
Francie Anne Riley

By overlaying stamps, you get the natural look of fallen leaves on a forest floor.

Floorcloth size: 2' x 3'

GATHER THESE SUPPLIES

Painting Surface:
Canvas or vinyl remnant,
    measuring 2' x 3'

Paints & Coatings:
Indoor/outdoor satin acrylics:
    Black
    Cumin
    Espresso
    Fairway Green
    Mint
    Mustard
    Olive

Tools:
Chalk pencil
Paint roller
Sponging mitt or sea sponge
Tape measure

Other Supplies:
Disposable plate
Masking tape: 1"-wide
Sponge
Stamps: Leaf Collection

INSTRUCTIONS

Prepare:
1. Refer to Preparing a Floorcloth on page 15. Measure, cut, and prepare floorcloth.

Continued on page 111

Forest Floor
Instructions begin
on page 109

Continued from page 109

2. Using chalk pencil and tape measure, mark 2½" to 3" from the edges for the border. Mask inside border.

Paint the Design:
1. Using a damp sponging mitt, pounce on Cumin, Espresso, and Mustard in border area. Using sponge, mottle colors. Remove tape and let dry.

2. Reposition tape on border around edge of inside area.

3. Using paint roller, apply one coat of Mint. This will look bright but, when softened with other leaf colors, it will offer needed highlights. Coverage does not need to be opaque. Be certain entire area is covered since Mint serves as the primer/sealer for the surface. Let dry.

4. Using clean, moistened sponge, lightly pounce on mixture of Cumin and Olive. Do not cover area completely. Allow Mint to show through. Let dry several hours before stamping.

5. Using disposable plate, mix two parts Fairway Green with half part Black. Use this mix and Olive Green to stamp leaves on surface.

6. Using Leaf Collection designs, stamp leaves at random around floorcloth.

7. Alternate types of leaves as well as green shades. Mix greens with Cumin on some pressings.

8. After second pressing with each paint loading, position leaf to do a light pressing only in areas where primary pressings have dried. Let dry.

9. Add additional light pressings with Cumin to look faded in background.

Finish:
1. Let floorcloth dry 72 hours. With indoor/outdoor satin acrylics, there is no need to seal floorcloth.✳

Tulips & Ferns

Pictured on page 112

Designed by
Kathi Malarchuk

This combination of elegant stamps and rich colors creates a stunning floorcloth.

Floorcloth size: 4' x 5½'

GATHER THESE SUPPLIES

Painting Surface:
Canvas or vinyl remnant, measuring 4' x 5½'

Paints & Coatings:
Faux-finishing glazes:
    Bark Brown
    Neutral
Indoor/outdoor satin acrylics:
    Barn Red
    Black
    Eggplant
    Mustard
    Olive
Matte acrylic sealer

Tools:
Chalk pencil
Glaze roller
Paint roller
Sponging mitt or sea sponge
Tape measure

Other Supplies:
Disposable plate
Geometric background stencil:
    Octagon
Masking tape: 1"-wide
Stamps:
    Blooming Vine
    Martha's Fern
    Tulip

INSTRUCTIONS

Prepare:
1. Refer to Preparing a Floorcloth on page 15. Measure, cut, and prepare floorcloth.

2. Using paint roller, base-coat floorcloth with Black. Let cure for 72 hours.

Paint the Design:
1. Using chalk pencil and tape measure, mark in 3" from all edges for outer border. Mask along inside edge of border.

2. Measure in 8" from outer edge and mark 1½"-wide inner border. Mask on both sides of this border.

3. Using glaze roller, paint borders with Barn Red. Remove tape and let dry.

4. Measure in 4" from outer edge and mark for 1"-wide border inside outer border. Mask along both edges of this border.

Continued on page 113

Tulips & Ferns
Instructions begin
on page 111

Continued from page 111

5. Repeat at 7" from outer edge for another border around outside of inner Barn Red border.

6. Using glaze roller, paint borders with Mustard. Remove tape and let dry.

7. Measure and mark vertical and horizontal center lines.

8. Align Octagon template so an octagon shape tile is in exact center of floorcloth. Paint octagon shape with Barn Red.

9. Repeat to paint all octagons within inner border. Let dry.

10. Using Blooming Vine design, stamp corners with Eggplant and Olive.

11. Stamp Tulips and Martha's Fern with Eggplant and Olive.

12. Let dry two to four hours.

13. Using disposable plate, blend Bark Brown glaze with Neutral glaze to achieve desired tint. Dampen a sponging mitt or sea sponge and dip in Bark Brown mixture, lightly sponging over entire floorcloth to give an aged look.

Finish:
1. Let floorcloth dry 72 hours. With indoor/outdoor satin acrylics, there is no need to seal floorcloth.✻

Bees & Dragonflies

Pictured on page 114

Designed by
Julie Watkins Schreiner

When the textures of squares, curves, and insects are combined, this flower-shaped floorcloth becomes alive with activity .

Floorcloth size: 36" x 36"

GATHER THESE SUPPLIES

Painting Surface:
Canvas or vinyl remnant, measuring 36" x 36"

Paints & Coatings:
Indoor/outdoor satin acrylics:
  Apricot
  Damask Blue
  Green Mist
  Khaki
  Light Blue
  Mojave Sunset
  Vanilla

Tools:
Craft knife
Craft scissors
Glaze roller
Pencil
Round brush: 1"

Other Supplies:
Geometric background stencil:
  Squares
Plate: 11", round (optional)
Posterboard
Stamps:
  Bee
  Dragonfly
Tracing paper & pencil
Transfer paper & stylus

INSTRUCTIONS

Prepare:
1. Refer to Preparing a Floorcloth on page 15. Measure, cut, and prepare floorcloth.

2. Using tracing and transfer tools, transfer Bees & Dragonflies Pattern on page 115 onto floorcloth. Using craft knife, cut out flower shape.

3. Using 11" plate, transfer circle in center of flower.

Paint the Design:
1. Using 1" round brush, paint circle with Green Mist. Let dry.

2. Using posterboard and scissors, cut 11" circle to mask out flower center.

3. Using Squares design, stencil squares alternately with Apricot, Damask Blue, Khaki, Light Blue, and Mojave Sunset. Refer to photo on page 114 for color placement. Let dry thoroughly before continuing.

4. Using Bee and Dragonfly designs, randomly stamp insects around floorcloth, some with Khaki and some with Vanilla. Use edge of posterboard to mask off edge of adjacent square on some of motifs which are stamped between two squares to create the illusion of insects going under squares. Let others overlap onto adjacent square.

5. Using 1" round brush, lightly freehand a spiral in center of circle with Damask Blue, starting from a Damask Blue square.

Continued on page 115

Bees &
Dragonflies
Instructions begin
on page 113

Bees & Dragonflies Pattern

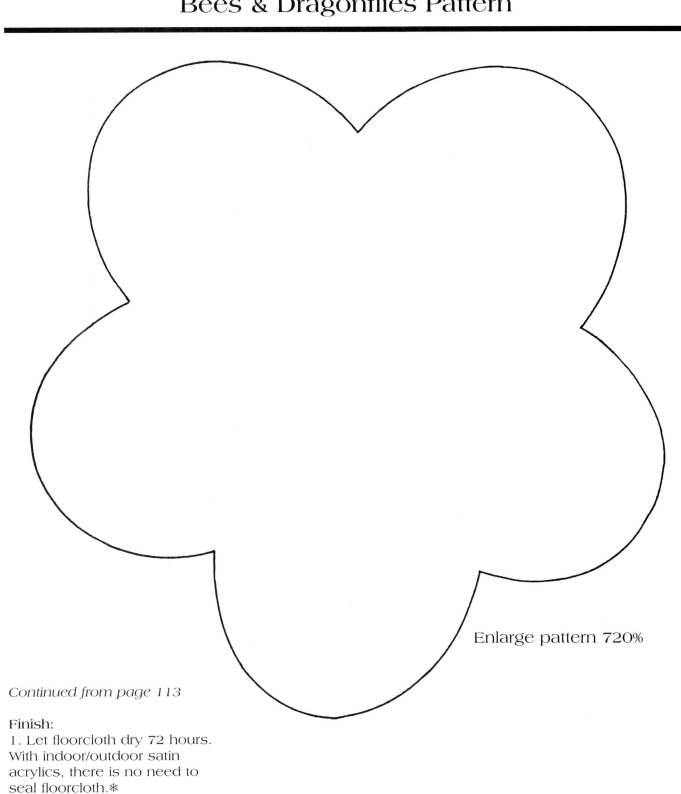

Enlarge pattern 720%

Continued from page 113

Finish:
1. Let floorcloth dry 72 hours. With indoor/outdoor satin acrylics, there is no need to seal floorcloth.✳

Garden
Reflecting Pond
Instructions begin
on page 117

Garden Reflecting Pond

Pictured on page 116

Based on a design © by
Kiki Farish

This floorcloth is a great example of random design. Have fun by creating your own vision of a pond brimming with creatures.

Floorcloth size: 2' x 3'

GATHER THESE SUPPLIES

Painting Surface:
Canvas or vinyl remnant,
    measuring 2' x 3'

Paints & Coatings:
Faux-finishing glaze:
    Penny Copper
Indoor/outdoor satin acrylics:
    Cumin
    Damask Blue
    Fairway Green
    Green Mist
    Ink Blue
    Mint
    Vanilla
    White
Matte acrylic sealer

Tools:
Paint roller
Pencil
Sponge brush: 2"
Sponging mitt or sea sponge

Other Supplies:
Butcher paper: 2⅛' x 3⅛'
Disposable plate
Masking tape
Paper towels

Stamps:
    Atlantic Star
    Bee
    Classic Koi
    Dragonfly
    Martha's Fern
    Nautilus

INSTRUCTIONS

Prepare:
1. Refer to Preparing a Floorcloth on page 15. Measure, cut, and prepare floorcloth.

2. Using paint roller, base-coat floorcloth with Mint. Coverage does not need to be opaque. Any slight areas of background surface showing through will add to texture of finished "pond."

Paint the Design:
1. Using pencil, mark outside edges of floorcloth on butcher paper. Sketch in size and shape of "pond." Make this irregular rather than perfect oval or circle.

2. Tear paper as close as possible to sketched pond line. Keep both "inner pond" area and outer "bank" area of paper to use as masks.

3. Mask inner pond piece onto floorcloth. This becomes a cover while sponging bank area.

4. Lightly moisten sponging mitt with water and remove excess water with paper towel. Dab one edge of sponge in Damask Blue and another edge in Ink Blue. Lightly pounce from edge of paper to outside edge of floorcloth. Add the greatest amount of color around pond's edge, making it lighter while moving to outer edge. Refer to photo on page 116 for placement. Let dry.

5. Remove "pond" paper and position "bank" section of paper over what has been sponged. This will expose pond area.

6. Using disposable plate, mix equal parts Vanilla and White, but do not mix together completely. Dip sponge brush in water then pick up some blended paint. Brush "wet" across pond to give a watery look. Let dry.

7. Using Atlantic Star, Bee, Classic Koi, Dragonfly, Martha's Fern, and Nautilus designs, stamp randomly on pond area with Fairway Green, Green Mist, and White. Let dry.

8. Brush over some designs with watery Vanilla mixed with White to make them appear deeper in pond. Let dry.

9. Add additional light pressings with Cumin and Penny Copper glaze so they look faded into background.

Finish:
1. Let floorcloth dry 72 hours. With indoor/outdoor satin acrylics, there is no need to seal floorcloth.✳

Zebra Skin Rug
Instructions begin
on page 119

Zebra Skin Rug

Pictured on page 118

Designed by
Kathi Malarchuk

You can liven up any room in your home with this fun, animal-inspired floorcloth.

Floorcloth size: 4' x 6'

GATHER THESE SUPPLIES

Painting Surface:
Canvas or vinyl remnant,
 measuring 4' x 6'

Paints & Coatings:
Indoor/outdoor satin acrylics:
 Caramel
 Oxblood

Tools:
Chalk pencil
Paint roller
Tape measure

Other Supplies:
Geometric background stencil:
 Diamond
Masking tape: 1"-wide
Stamps: Safari Kit

INSTRUCTIONS

Prepare:
1. Refer to Preparing a Floorcloth on page 15. Measure, cut, and prepare floorcloth.

2. Using paint roller, base-coat floorcloth with Caramel. Let cure for 72 hours.

3. Using chalk pencil and tape measure, mark vertical and horizontal center lines of floorcloth.

4. Using Diamond stencil, align template A so a complete diamond is in the exact center of floorcloth. Mark diamond alignment corners and repeat for entire floorcloth.

5. Align template B in center. Using animal skin design, stamp with Oxblood. Repeat stamping to fill entire diamond. Repeat to complete all the diamonds on the entire rug. Let dry 24 hours.

6. Mark in 2" from outer edges creating inner edge of outer border.

7. Mark in 3½" from last marking (5½" from outer edge) creating outer edge of inner border.

8. Inner border is 3½" wide. Mask along inside of outer border and on both sides of inner border.

Paint the Design:
1. Using paint roller, paint borders with Oxblood. Remove tape carefully. Let dry.

2. Using zebra design and glaze roller, stamp on inner border with Caramel. Refer to photo on page 118 for placement.

Finish:
1. Let floorcloth dry 72 hours. With indoor/outdoor satin acrylics, there is no need to seal floorcloth.✳

Summer Fruits
Instructions begin
on page 121

Summer Fruits

Pictured on page 120

Based on a design © by
Rhonda Kaplan

This "dropped-in rug" was the perfect solution for repairing a piece of vinyl flooring. This technique shows that you can paint directly on top of inlaid flooring. If you choose, you could also create a floorcloth with this design.

Floorcloth size: existing inlaid
   flooring

GATHER THESE SUPPLIES

Painting Surface:
Existing inlaid flooring

Paints & Coatings:
Indoor/outdoor satin acrylics:
   Barn Red
   Black
   Fairway Green
   Mint
   Mustard
   White

Tools:
Flat brush: 2"
Paint roller
Sponge wedges
Stencil brush: 1"

Other Supplies:
Grout tape, ¼"
Masking tape: 1"-wide
Sandpaper: #100
Single-overlay stencil:
   Checkerboard Collection
Stamps:
   Fruit Decor Kit
   Grape Cluster

INSTRUCTIONS

Prepare:
1. Refer to Preparing a Floorcloth on page 15. Remove all wax or dirt buildup. Thoroughly rinse all cleaning chemicals from floor surface. Let dry.

2. Using 1"-wide masking tape, mask off floor around outer edge of area to be painted.

3. Using sandpaper, sand area inside tape to remove top coating. Thoroughly remove all dust and residue.

Paint the Design:
1. Using paint roller, paint area with two coats of White. Let dry.

2. Using Checkerboard stencil, mark a three-row checkerboard around outer edge of area.

3. Mask across stencil as needed to create diagonal, inside corners of area inside checkerboard.

4. Using 1" stencil brush, stencil checks with Black. Use a pouncing stroke to get a solid buildup of paint. Let dry.

5. Using 2" flat brush, paint some white squares of the checkerboard with Mint and Mustard.

6. Measure and mark center section horizontally and vertically to create quarters.

7. Mask center area to protect checkerboard border.

8. Mask over horizontal and vertical lines that divide center area, with grout tape. These areas will remain White.

9. Using glaze roller, paint two opposite quarters with Black, another quarter with Mustard, and remaining quarter with Mint. Let dry.

10. Using Fruit Decor Kit designs and Grape Cluster design, stamp all fruit motifs with White. Let dry.

11. Reload stamps with your choice of Barn Red, Fairway Green, Mint, or Mustard and stamp over White to give added dimension to each print. Refer to photo on page 120 for color placement.

12. Using paint roller, apply base color to stamp. Using sponge wedges, pounce on a shadow color.

13. Using 2" flat brush, shadow outer edge of pears with inky Black.

14. Remove tape.

Finish:
1. Let cure for 72 hours. If repair is in a high traffic area, protect surface for several more days by placing several brown paper bags on painted surface and walking gently over bags until paint has had time to thoroughly cure.✳

Abundant Fruit
Instructions begin
on page 123

Abundant Fruit

Pictured on page 122

Designed by
Kathi Malarchuk

This fruit-filled floorcloth is a nice combination of symmetrical borders and random interior design. It would look great in any kitchen.

Floorcloth size: 3' x 4'

GATHER THESE SUPPLIES

Painting Surface:
Canvas or vinyl remnant,
    measuring 3' x 4'

Paints & Coatings:
Faux-finishing glazes:
    Blue Iris
    Cerulean
    Neutral
    White

Tools:
Chalk pencil
Paint roller
Stippler brush
Tape measure

Other Supplies:
Blocks:
    Casablanca Medallion
    Fruits
Disposable plate
Masking tape: 1"-wide

INSTRUCTIONS

Prepare:
1. Refer to Preparing a Floor-cloth on page 15. Measure, cut, and prepare floorcloth.

2. Using chalk pencil and tape measure, mark 3"-wide border around outer edge and mark. Measure in 4", leaving white space inside border 4"-wide and mark. Mask along marks. Mask off border 2"-wide inside center.

Paint the Design:
1. Using disposable plate, mix small part Neutral with Blue Iris. Using stippler brush, stipple borders with glaze mixture. Remove tape. Let dry.

2. Mask off around inner border. Using paint roller, paint center area of border with Cerulean. Remove tape. Let dry.

3. Using Casablanca Medallion design, stamp in white area of floorcloth with Cerulean.

4. Using Fruits design, stamp fruits in center area with Blue Iris, Cerulean, and White. Let dry.

Finish:
1. With indoor/outdoor satin acrylics, there is no need to seal floorcloth.✳

Vineyard Delight

Pictured on page 124

Designed by
Kathi Malarchuk

This floorcloth uses a lattice background to accentuate grapes and grape leaves for an overall, bright and clean look.

Floorcloth size: 3' x 5'

GATHER THESE SUPPLIES

Painting Surface:
Canvas or vinyl remnant,
    measuring 3' x 5'

Paints & Coatings:
Indoor/outdoor satin acrylics:
    Damask Blue
    Green Mist
    Mint
    White

Tools:
Chalk pencil
Paint roller
Stippler brush
Tape measure

Other Supplies:
Disposable plate
Geometric background stencil:
    Lattice
Masking tape: 1"-wide
Paper towels
Stamps: Grape Collection

INSTRUCTIONS

Prepare:
1. Refer to Preparing a Floor-cloth on page 15. Measure, cut, and prepare floorcloth.

2. Using paint roller, base-coat floorcloth with Green Mist. Let dry.

3. Using chalk pencil and tape measure, mark 2"-wide border around entire edge of floor-cloth. Place tape along inside of this line.

4. Mask 10" in from edge along both sides of 1"-wide inner border.

Paint the Design:
1. Using paint roller, paint borders with Damask Blue. Remove tape. Let dry.

Continued on page 125

Vineyard Delight
Instructions begin
on page 123

Continued from page 123

2. Mask off area around inside inner border, masking on painted border.

3. Pour small part White onto disposable plate. Dip stippler brush into paint and wipe off excess on paper towel. Lightly stipple area inside inner border.

4. Using Lattice design, stencil trellis around outer area (between inside and outside borders) with Damask Blue and Mint. Refer to photo on page 124 for placement.

5. Using Grape Collection designs, stamp on inner stippled area with Damask Blue. Stamp all leaves except center leaves with Mint. Stamp center leaves with White. Let dry.

Finish:
1. With indoor/outdoor satin acrylics, there is no need to seal floorcloth.✳

Metric Conversion Chart

INCHES	MM	CM	INCHES	CM	INCHES	CM
⅛	3	0.3	9	22.9	30	76.2
¼	6	0.6	10	25.4	31	78.7
½	13	1.3	12	30.5	33	83.8
⅝	16	1.6	13	33.0	34	86.4
¾	19	1.9	14	35.6	35	88.9
⅞	22	2.2	15	38.1	36	91.4
1	25	2.5	16	40.6	37	94.0
1¼	32	3.2	17	43.2	38	96.5
1½	38	3.8	18	45.7	39	99.1
1¾	44	4.4	19	48.3	40	101.6
2	51	5.1	20	50.8	41	104.1
2½	64	6.4	21	53.3	42	106.7
3	76	7.6	22	55.9	43	109.2
3½	89	8.9	23	58.4	44	111.8
4	102	10.2	24	61.0	45	114.3
4½	114	11.4	25	63.5	46	116.8
5	127	12.7	26	66.0	47	119.4
6	152	15.2	27	68.6	48	121.9
7	178	17.8	28	71.1	49	124.5
8	203	20.3	29	73.7	50	127.0

Product Sources

The products used to create the floorcloths in this book can be found at craft and hobby stores. Below is a listing of brand names used for the projects.

Decorator Products®

This group of products encompasses colored glazes, indoor/outdoor satin acrylics, printing blocks and stamps, decorator tools, geometric background stencils, and faux-finishing kits.

Decorator Blocks® used for Block Printing:
Fruits #53252
Ivy #53202
Little Garden Flowers #53219

Decorator Blocks® Stamp Decor™ used for Stamping:
Acanthus Leaf #53609
Atlantic Star #53624
Bee #53620
Blooming Vine #53605
Casablanca #53601
Classic Koi #53621
Dragonfly #53619
Fruits Collection #53669
Grape Cluster #53614
Grapes Collection #53664
Leaves Collection #53661
Magnolias Collection #53663
Martha's Fern #53610
Nautilus #53625
Safari Kit #53668
Tulip #53618

Foundations™ Wall & Floor Templates:
Used for creating geometric background patterns on walls and floors.
Damask #53705
Diamond #53701
Lattice #53703
Octagon #53704
Squares #53705

Decorator Glazes™:
Used for stamping and block printing as well as mixing with neutral glazing medium for texturing.
Alpine Green #53044
Baby Pink #53015
Bark Brown #53033
Blue Bell #53023
Blue Iris #53054
Cerulean #53056
Danish Blue #53024
Deep Woods Green #53032
Italian Sage #53045
Ivy Green #53031
Moss Green #53043
Mushroom #53049
Neutral #53001
Penny Copper #53005
Persimmon #53010
Russet Brown #53050
Sage Green #53028
Sunflower #53008
White #53006

Durable Colors™:
Used as an indoor/outdoor satin acrylic when stenciling, stamping, and block printing.
Apricot #53305
Barn Red #53308
Black #53328
Caramel #53324
Chamois #53303
Cumin #53325
Damask Blue #53314
Eggplant #53312
Espresso #53327
Fairway Green #53319
Green Mist #53318
Ink Blue #53316
Khaki #53323
Light Blue #53313
Limestone #53321
Mint #53317
Mojave Sunset #53306
Mustard #53304
Olive #53320
Oxblood #53310

Pediment #53322
Twig #53326
Vanilla #53302
White #53301

Decorator Products® Tools:
Chamois Tool #30130
Decorator Blocks® Brush Set #53453
Flogger #30115
French Brush #30122
Glaze Applicators #53477
Glaze Roller #30118
Grout Tape #30117
Marbleizing Feathers #30101
Marbleizing Sponge #30102
Mopping Mitt #30107
Ragging Mitt #30106
Spatter Tool #30121
Sponging Mitt #30105
Stamp Decor™ Stamp Cleaner #53478
Stippler Brush #30128

Decorator Products® Faux-Finish Kit:
Splashed Marble #30080

Decorator Products® Decorator Sealers:
Aerosol Matte Sealer #30144
Aerosol Thick Gloss Sealer #30143
Liquid Gloss Sealer #30141
Liquid Matte Sealer #30142

FolkArt® Acrylic Colors:
These are high-quality brush-on acrylic paints. They are premixed colors available in 170 color hues. Brushes clean up with soap and water.
Berry Wine #434
Bright Peach #682
Burgundy #957
Butter Pecan #939
Buttercrunch #737
Clay Bisque #601
Dapple Gray #937

Dove Gray #708
Grass Green #644
Green Forest #448
Green Meadow #726
Holiday Red #612
Licorice #938
Light Gray #424
Maple Syrup #945
Midnight #964
Mint Green #445
Mystic Green #723
Poetry Green #619
Raspberry Sherbet #966
Raspberry Wine #935
Rose Chiffon #753
Skintone #949
Slate Blue #910
Spring White #430
Sunflower #432
Tapioca #903
Tartan Green #725
Teal #405
Teddy Bear Tan #419
Wicker White #901
Winter White #429

FolkArt® Metallic Colors:
Inca Gold #676
Pure Gold #660

FolkArt® Artists' Pigments:
Asphaltum #476
Burnt Umber #462
Cobalt #720
Dioxazine Purple #463
Hauser Green Dark #461
Medium Yellow #455
Napthol Crimson #435
Sap Green #458
Turner's Yellow #679
Warm White #649

FolkArt® Mediums:
Blending Gel Medium
Extender
Thickener

FolkArt® Finishes:
Outdoor Sealer
Available in gloss, matte, or
satin finishes.
Clearcote Extra Thick Glaze
Clearcote Hi-Shine Glaze:
Clearcote Matte Acrylic Sealer

FolkArt® One Stroke™ Brushes:
Mini Scruffy Brush #1174
Scruffy Brush #1172

Stencil Decor® Products
Dry Brush™ Stencil Paints
Andiron Black #26224
Berry Red #26241
Cameo Peach #26201
Ecru Lace #26216
English Lavender #26217
Forest Shade #26239
Gold Metallic #26234
Herb Garden Green #26208
Ol' Pioneer Red #26211
Promenade Rose #26204
Romantic Rose #26213
Sage Green #26229
Shadow Gray #26237
Sherwood Forest Green #26209
Ship's Fleet Navy #26207
Sunny Brooke Yellow #26218
Terra Cotta #26240
True Blue #26232
Truffles Brown #26206
Wild Ivy Green #26210
Wildflower Honey #262157

Elegant Home™ Gel Paints:
These gel formula paints are
used to enhance detail of stencils.
Cactus #26125
Daffodil Yellow #26104
Forest Shade #26121
Ivory Lace #26102
King's Gold #26132
Napa Grape #26112
Pink Blush #26109
Shadow Gray #26131
Tempest Blue #26117
Twig #26128
Wedgwood Blue #26116
Wild Ivy #26126
Wood Rose #26110

Stencil Decor® Stencils:
Birdhouses #26853
Gardener's Ivy #26620
Nature's Vineyard #26614
Noah's Ark #26701
Pots and Planters #26854
Rose Swag #26602
Uncut Stencil Blank #26667
Victoria's Rose #26669

Elegant Home™ Stencils:
Apple Blossom & Bird's Nest
#29006
Ginger Jar & Fern #29008
Ivy #29002

Simply® Stencils:
Checkerboard #28444
Checkerboard Collection #28771
Classic Ivy #28901
Color Crayons #28454
Cottage Rose #28573
Country Greetings #28559
Fruit Medley #28367
Greek Key #28148
Up, Up and Away #28347

Stencil Decor® Designer Tape™:
Basket Weave, 2" #26402

Stencil Decor® Tools:
Spouncer™ sponge pouncing tool
Stencil Brush Cleaner #26251
Stencil Brushes
Stencil Roller #34006
Stencil Tape #34002

Apple Barrel® Acrylic Gloss
Enamels:
Used as indoor/outdoor gloss
enamels and available in 40
decorator colors.
Dandelion Yellow #20646
Deep Purple #20625
Raspberry #20629
Real Green #20651
Real Red #20636
Spring Green #20652
White #20621

Royal Coat® Decoupage Finish:
Glues and seals paper and fabric
for découpaging.

Index